AF522312

# BIOINFORMATICS AND MOLECULAR EVOLUTION

# BIOINFORMATICS AND MOLECULAR EVOLUTION

Naresh Mahajan

RANDOM PUBLICATIONS

NEW DELHI - 110 002 (INDIA)

**Bioinformatics and Molecular Evolution**

---

ISBN 978-93-51116-93-6



Published in 2015 in India by
Reprint 2019

**RANDOM PUBLICATIONS**

4376-A/4B, Gali Murari Lal, Ansari Road
New Delhi-110 002
Phone: +9111-43580356, 23289044
E-mail: randomexports@gmail.com; sales@randompublications.com;
info@randompublications.com

*Type Setting by:* Friends Media, Delhi-110089
*Printed at :* Mehra Printers, Delhi-110092

# Preface

For evolutionists the revolution in DNA technology has been a major advance. The reason is that the very nature of DNA allows it to be used as a "document" of evolutionary history: comparisons of the DNA sequences of various genes between different organisms can tell us a lot about the relationships of organisms that cannot be correctly inferred from morphology. One definite problem is that the DNA itself is a scattered and fragmentary "document" of history and we have to beware of the effects of changes in the genome that can bias our picture of organismal evolution. Molecular evolution is a change in the sequence composition of cellular molecules such as DNA, RNA, and proteins over long periods of time. Molecular evolution attempts to explain biological change at the molecular and cellular level using the principles of evolutionary biology and population genetics. Major topics in molecular evolution concern the rates and impacts of single nucleotide change, neutral evolution vs. natural selection, origins of new genes, the genetic nature of complex traits, the genetic basis of speciation, evolution of development, and ways that evolutionary forces influence genomic and phenotypic changes.

The content and structure of a genome is the product of the molecular and population genetic forces which act upon that genome. Novel genetic variants will arise through mutation and will spread and be maintained in populations due to genetic drift or natural selection. Genome size is influenced by the amount of repetitive DNA as well as number of genes in an organism. The C-value paradox refers to the lack of correlation between organism 'complexity' and genome size. Explanations for the so-called paradox are two-fold. First, repetitive genetic elements can comprise large portions of the genome for many organisms, thereby inflating DNA content of the haploid genome. Secondly, the number of genes is not necessarily indicative of the number of developmental stages or tissue types in an organism. An organism with few developmental stages or tissue types may have large numbers of genes that influence non-developmental phenotypes, inflating gene content relative to developmental gene families. Neutral explanations for genome size suggest that when population sizes are small, many mutations become nearly neutral. Hence, in small populations repetitive content and other 'junk' DNA can accumulate without placing the organism at a competitive disadvantage. There is little evidence to suggest that genome size is under strong widespread selection in multicellular eukaryotes. Genome size, independent of gene content, correlates poorly with most physiological traits and many eukaryotes, including mammals, harbour very large amounts of repetitive DNA. However, birds likely have experienced strong selection for reduced genome size, in response to changing energetic needs for flight. Birds, unlike humans, produce nucleated red blood cells, and larger nuclei lead to lower levels of oxygen transport. Bird metabolism is far higher than that of mammals, due largely to flight, and oxygen needs are high. Hence, most birds have small, compact genomes with few repetitive elements. Indirect evidence suggests that non-avian theropod dinosaur ancestors of modern birds also had reduced genome sizes, consistent with endothermy and high energetic needs for running speed. Many bacteria have also experienced selection for small genome size, as time of replication and energy consumption are so tightly correlated with fitness. Over the past few decades, major advances in the field of molecular biology, coupled with advances in genomic technologies, have led to an explosive growth in the biological information generated by researchers. This deluge of genomic information has, in turn, led to an absolute requirement for computerised databases to store, organise and index the data, and for specialised tools to view and analyse the data. The different kinds of structural analysis led to several different kinds of databases. The molecular biology have elongated in several different but strongly related working fields. Bioinformatics is an interdisciplinary research area that is the interface between the biological and computational sciences. The ultimate goal of bioinformatics is to uncover the wealth of biological information hidden in the mass of data and obtain a clearer insight into the fundamental biology of organisms. This new knowledge could have profound impacts on fields as varied as human health, agriculture, the environment, energy and biotechnology.

Molecular evolution is a very active research area. This book discusses the interface between statistics and molecular evolution.

I thank all members of my team who have helped in the preparation of the book. My special thanks go to "Random Publications" who have published the book.

*— Naresh Mahajan*

# Contents

# Chapter 1

# Introduction to Bioinformatics

Bioinformatics was applied in the creation and maintenance of a database to store biological information at the beginning of the "genomic revolution", such as nucleotide and amino acid sequences. Development of this type of database involved not only design issues but the development of complex interfaces whereby researchers could both access existing data as well as submit new or revised data. In order to study how normal cellular activities are altered in different disease states, the biological data must be combined to form a comprehensive picture of these activities. Therefore, the field of bioinformatics has evolved such that the most pressing task now involves the analysis and interpretation of various types of data, including nucleotide and amino acid sequences, protein domains, and protein structures. The actual process of analysing and interpreting data is referred to as computational biology. Important sub-disciplines within bioinformatics and computational biology include:

- the development and implementation of tools that enable efficient access to, and use and management of, various types of information.
- the development of new algorithms (mathematical formulas) and statistics with which to assess relationships among members of large data sets, such as methods to locate a gene within a sequence, predict protein structure and/or function, and cluster protein sequences into families of related sequences.

There are two fundamental ways of modelling a Biological system (e.g. living cell) both coming under Bioinformatic approaches;

- Static;
  - o Sequences-Proteins, Nucleic acids and Peptides

    - Structures-Proteins, Nucleic acids, Ligands (including metabolites and drugs) and Peptides
    - Interaction data among the above entities including microarray data and Networks of proteins, metabolites.
- Dynamic;
    - Systems Biology comes under this category including reaction fluxes and variable concentrations of metabolites
    - Multi-Agent Based modelling approaches capturing cellular events such as signalling, transcription and reaction dynamics.

A broad sub-category under bioinformatics is structural bioinformatics.

Structural bioinformatics is the branch of bioinformatics which is related to the analysis and prediction of the three-dimensional structure of biological macromolecules such as proteins, RNA, and DNA. It deals with generalizations about macromolecular 3D structure such as comparisons of overall folds and local motifs, principles of molecular folding, evolution, and binding interactions, and structure/function relationships, working both from experimentally solved structures and from computational models.

## Major Research Areas

Since the Phage Φ-X174 was sequenced in 1977, the DNA sequences of thousands of organisms have been decoded and stored in databases. This sequence information is analysed to determine genes that encode polypeptides (proteins), RNA genes, regulatory sequences, structural motifs, and repetitive sequences. A comparison of genes within a species or between different species can show similarities between protein functions, or relations between species (the use of molecular systematics to construct phylogenetic trees). With the growing amount of data, it long ago became impractical to analyse DNA sequences manually. Today, computer programs such as BLAST are used daily to search sequences from more than 260 000 organisms, containing over 190 billion nucleotides.

These programs can compensate for mutations (exchanged, deleted or inserted bases) in the DNA sequence, in order to identify sequences that are related, but not identical. A variant of this sequence alignment is used in the sequencing process itself. The so-called shotgun sequencing technique (which was used, for example, by The Institute for Genomic Research to sequence the first bacterial genome, *Haemophilus influenzae*) does not produce entire chromosomes, but instead generates the sequences of many thousands of small DNA fragments (ranging from 35 to 900 nucleotides long, depending on the sequencing technology).

The ends of these fragments overlap and, when aligned properly by a genome assembly program, can be used to reconstruct the complete genome. Shotgun sequencing yields sequence data quickly, but the task of assembling the fragments can be quite complicated for larger genomes. For a genome as large as the human genome, it may take many days of CPU time on large-memory, multiprocessor computers to assemble the fragments, and the resulting assembly will usually contain numerous gaps that have to be filled in later. Shotgun sequencing is the method of choice for virtually all genomes sequenced today, and genome assembly algorithms are a critical area of bioinformatics research. Another aspect of bioinformatics in sequence analysis is annotation, which involves computational gene finding to search for protein-coding genes, RNA genes, and other functional sequences within a genome. Not all of the nucleotides within a genome are part of genes. Within the genome of higher organisms, large parts of the DNA do not serve any obvious purpose. This so-called junk DNA may, however, contain unrecognized functional elements. Bioinformatics helps to bridge the gap between genome and proteome projects — for example, in the use of DNA sequences for protein identification.

## Sequence Analysis

The term "sequence analysis" in biology implies subjecting a DNA or peptide sequence to sequence alignment, sequence databases, repeated sequence searches, or other bioinformatics methods on a computer.

Since the development of methods of high-throughput production of gene and protein sequences during the 90s, the rate of addition of new sequences to the databases increases continuously. Such a collection of sequences does not, by itself, increase the scientist's understanding of the biology of organisms. However, comparing sequences with known functions with these new sequences is one way of understanding the biology of that organism from which the new sequence comes. Thus, sequence analysis can be used to assign function to genes and proteins by the study of the similarities between the compared sequences. Nowadays there are many tools and techniques that provide the sequence comparisons (sequence alignment) and analyse the alignment product to understand the biology.

Sequence analysis in molecular biology and bioinformatics is an automated, computer-based examination of characteristic fragments, e.g. of a DNA strand. It basically includes relevant topics:

1. The comparison of sequences in order to find similarity and dissimilarity in compared sequences (sequence alignment)
2. Identification of gene-structures, reading frames, distributions of introns and exons and regulatory elements
3. Finding and comparing point mutations or the single nucleotide polymorphism (SNP) in organism in order to get the genetic marker.
4. Revealing the evolution and genetic diversity of organisms.
5. Function annotation of genes.

In chemistry, sequence analysis comprises techniques used to do determine the sequence of a polymer formed of several monomers. In molecular biology and genetics, the same process is called simply "sequencing". In marketing, sequence analysis is often used in analytical customer relationship management applications, such as NPTB models.

### *Methodology*

For sequence alignment method compose of pairwise alignment (align with two sequences) and multiple alignment (align with more than two sequence). There are several tools for alignment, including: ClustalW, PROBCONS, MUSCLE, MAFFT, DIALIGN, T-Coffee, POA, and MANGO.

There are many algorithms used in the study of sequence analysis. These include: dynamic programming, Hidden Markov Model, Viterbi and Greedy.

## Sequence Profiling Tool

A sequence profiling tool in bioinformatics is a type of software that presents information related to a genetic sequence, gene name, or keyword input. Such tools generally take a query such as a DNA, RNA, or protein sequence or 'keyword' and search one or more databases for information related to that sequence. Summaries and aggregate results are provided in standardized format describing the information that would otherwise have required visits to many smaller sites or direct literature searches to compile. Many sequence profiling tools are software portals or gateways that simplify the process of finding information about a query in the large and growing number of bioinformatics databases. The access to these kinds of tools is either web based or locally downloadable executables.

### *Introduction and Usage*

The "post-genomics" era has given rise to a range of web-based tools and software to compile, organize, and deliver large amounts of primary sequence information, as well as protein structures, gene annotations, sequence alignments, and other common bioinformatics tasks.

In general, there exist three types of databases and service providers. The first one includes the popular public-domain or open-access databases supported by funding and grants such as NCBI, ExPASy, Ensembl, and PDB. The second one includes smaller or more specific databases organized and compiled by individual research groups Examples include Yeast Genome Database, RNA database. The third and final one includes private corporate or institutional databases that require payment or institutional affiliation to access. Such examples rare given the globalization of the public databases unless the purported service is 'in-development' or the end point of the analysis is of commercial value.

Typical scenarios of a profiling approach become relevant, particularly, in the cases of the first two groups, where researchers commonly wish to combine information derived from several sources about a single query or target sequence. For example, users might use the sequence alignment and search tool BLAST to identify homologs of their gene of interest in other species, and then use these results to locate a solved protein structure for one of the homologs. Similarly, they might also want to know the likely secondary structure of the mRNA encoding the gene of interest, or whether a company sells a DNA construct containing the gene. Sequence profiling tools serve to automate and integrate the process of seeking such disparate information by rendering the process of searching several different external databases transparent to the user.

Many public databases are already extensively linked so that complementary information in another database is easily accessible; for example, Genbank and the PDB are closely intertwined. However, specialized tools organized and hosted by specific research groups can be difficult to integrate into this linkage effort because they are narrowly focused, are frequently modified, or use custom versions of common file formats. Advantages of sequence profiling tools include the ability to use multiple of these specialized tools in a single query and present the output with a common interface, the ability to direct the output of one set of tools or database searches into the input of another, and the capacity to disseminate hosting and compilation obligations to a network of research groups and institutions rather than a single centralized repository.

### *Keyword based Profilers*

Most of the profiling tools available on the web today fall into this category. The user, upon visiting the site/tool, enters any relevant information like a keyword e.g. dystrophy, diabetes etc., or GenBank accession numbers, PDB ID. All the relevant hits by the search are presented in a format unique to each tool's main focus. Profiling tools based on keyword searches are essentially search engines that are highly specialized for bioinformatics work, thereby eliminating a clutter of irrelevant or non-scholarly hits that might occur with a traditional search engine like Google. Most keyword-based profiling tools allow flexible types of keyword input, accession numbers from indexed databases as well as traditional keyword descriptors. Each profiling tool has its own focus and area of interest. For example, the NCBI search engine Entrez segregates its hits by category, so that users looking for protein structure information can screen out sequences with no corresponding structure, while users interested in perusing the literature on a subject can view abstracts of papers published in scholarly journals without distraction from gene or sequence results. The Pubmed biosciences literature database is a popular tool for literature searches, though this service is nearly equalled with the more general Google Scholar.

Keyword-based data aggregation services like the Bioinformatic Harvester performs provide reports from a variety of third-party servers in an *as-is* format so that users need not visit the website or install the software for each individual component service.

This is particularly invaluable given the rapid emergence of various sites providing different sequence analysis and manipulation tools. Another aggregative web portal, the Human Protein Reference Database (Hprd), contains manually annotated and curated entries for human proteins. The information provided is thus both selective and comprehensive, and the query format is flexible and intuitive. The pros of developing manually curated databases include presentation of proofread material and the concept of 'molecule authorities' to undertake the responsibility of specific proteins. However, the cons are that they are typically slower to update and may not contain very new or disputed data.

### *Sequence Data based Profilers*

A typical sequence profiling tool carries this further by using an actual DNA, RNA, or protein sequence as an input and allows the user to visit different web-based analysis tools to obtain the information desired. Such tools are also commonly supplied with commercial laboratory equipment like gene sequencers or sometimes sold as software applications for molecular biology. In another public-database example, the BLAST sequence search report from NCBI provides a link from its alignment report to other relevant information in its own databases, if such specific information exists.

For example, a retrieved record that contains a human sequence will carry a separate link that connects to its location on a human genome map; a record that contains a sequence for which a 3-D structure has been solved would carry a link that connects it to its structure database. Sequerome, a public service tool, links the entire BLAST report to many third party servers/sites that provide highly specific services in sequence manipulations such as restriction enzyme maps, open reading frame analyses for nucleotide sequences, and secondary structure prediction. The tool provides added advantage of maintaining a research log of the operations performed by the user, which can be then conveniently archived using 'mail', 'print' or 'save' functionality.

Thus an entire operation of researching on a sequence using different research tools and thus carrying a project to its completion can be completed within one browser interface. Consequently, future generation of sequence profiling tools would include ability to collaborate online with researchers to share project logs and research tools, annotate results of sequence analysis or lab work, customize and automate the processing of sets of sequence data etc. InstaSeq is a Google powered search tool that allows the user to directly enter a sequence and search the entire World Wide Web. This unique search engine, which is the only one of its kind, is in contrast to searching specific databases e.g. GenBank.

As a result the user can end up with a privately hosted document or a page from a lesser known database from just about anywhere in the world. Though the presence of sequence based profilers are far and few in the present scenario, their key role will become evident when huge amounts of sequence data need to be cross processed across portals and domains.

### Future Growth and Directions

The proliferation of bioinformatics tools for genetic analysis aids researchers in identifying and categorizing genes and gene sets of interest in their work; however, the large variety of tools that perform substantially similar aggregative and analytical functions can also confuse and frustrate new users. The decentralization encouraged by aggregative tools allows individual research groups to maintain specialized servers dedicated to specific types of data analysis in the expectation that their output will be collected into a larger report on a gene or protein of interest to other researchers. Data produced by microarray experiments, two-hybrid screening, and other high-throughput biological experiments is voluminous and difficult to analyse by hand; the efforts of structural genomics collaborations that are aimed at quickly solving large numbers of highly varied protein structures also increase the need for integration between sequence and structure databases and portals. This impetus toward developing more comprehensive and more user-friendly methods of sequence profiling makes this an active area of research among current genomics researchers.

## Sequence Motif

In genetics, a sequence motif is a nucleotide or amino-acid sequence pattern that is widespread and has, or is conjectured to have, a biological significance. For proteins, a sequence motif is distinguished from a structural motif, a motif formed by the three dimensional arrangement of amino acids, which may not be adjacent.

An example is the *N*-glycosylation site motif:

*Asn, followed by anything but Pro, followed by either Ser or Thr, followed by anything but Pro.*

where the three-letter abbreviations are the conventional designations for amino acids.

### *Overview*

When a sequence motif appears in the exon of a gene, it may encode the "structural motif" of a protein; that is a stereotypical element of the overall structure of the protein. Nevertheless, motifs need not be associated with a distinctive secondary structure. "Noncoding" sequences are not translated into proteins, and nucleic acids with such motifs need not deviate from the typical shape (e.g. the "B-form" DNA double helix).

Outside of gene exons, there exist regulatory sequence motifs and motifs within the "junk," such as satellite DNA. Some of these are believed to affect the shape of nucleic acids, but this is only sometimes the case. For example, many DNA binding proteins that have affinities for specific motifs only bind DNA in its double-helical form. They are able to recognize motifs through contact with the double helix's major or minor groove. Short coding motifs, which appear to lack secondary structure, include those that label proteins for delivery to particular parts of a cell, or mark them for phosphorylation. Within a sequence or database of sequences, researchers search and find motifs using computer-based techniques of sequence analysis, such as BLAST. Such techniques belong to the discipline of bioinformatics.

### *Motif Bioinformatics*

Consider the *N*-glycosylation site motif mentioned above:

*Asn, followed by anything but Pro, followed by either Ser or Thr, followed by anything but Pr.*

This pattern may be written as N{P}[ST]{P} where N = Asn, P = Pro, S = Ser, T = Thr; {X} means any amino acid except X; and [XY] means either X or Y.

The notation [XY] does not give any indication of the probability of X or Y occurring in the pattern. Sometimes patterns are defined in terms of a probabilistic model such as a hidden Markov model.

### *Motifs and Consensus Sequences*

The notation [XYZ] means X or Y or Z, but does not indicate the likelihood of any particular match. For this reason, two or more patterns are often associated with a single motif: the defining pattern, and various typical patterns. For example, the defining sequence for the IQ motif may be taken to be:

[FILV]Qxxx[RK]Gxxx[RK]xx[FILVWY]

where x signifies any amino acid, and the square brackets indicate an alternative.

Usually, however, the first letter is I, and both [RK] choices resolve to R. Since the last choice is so wide, the pattern IQxxxRGxxxR is sometimes equated with the IQ motif itself, but a more accurate description would be a *consensus sequence for the IQ motif.*

### *De Novo Computational Discovery of Motifs*

There are software programs which, given multiple input sequences, attempt to identify one or more candidate motifs. One example is MEME, which generates statistical information for each candidate. Other algorithms include CisModule, AlignAce, PhyloGibbs, Weeder, Amadeus and FIRE. SCOPE is an ensemble motif finder that uses several algorithms simultaneously. There currently exist more than 100 publications with similar algorithms without a comprehensive benchmark so selecting one is not straightforward.

### Discovery through Evolutionary Conservation

Motifs have been discovered by studying similar genes in different species. For example, by aligning the amino acid sequences specified by the GCM (*glial cells missing*) gene in man, mouse and *D. melanogaster*, Akiyama and others discovered a pattern which they called the GCM motif. It spans about 150 amino acid residues, and begins as follows:

WDIND*.*P..*...D.F.*W***.**.IYS**...A.*H*S*WAMRNTNNHN

Here each. signifies a single amino acid or a gap, and each * indicates one member of a closely-related family of amino acids.

The authors were able to show that the motif has DNA binding activity. A motif discovery algorithm that considers phylogenetic conservation is PhyloGibbs.

### Pattern Description Notations

Several notations for describing motifs are in use but most of them are variants of standard notations for regular expressions and use these conventions:

- there is an alphabet of single characters, each denoting a specific amino acid or a set of amino acids;
- a string of characters drawn from the alphabet denotes a sequence of the corresponding amino acids;
- any string of characters drawn from the alphabet enclosed in square brackets matches any one of the corresponding amino acids; e.g. [abc] matches any of the amino acids represented by a or b or c.

The fundamental idea behind all these notations is the matching principle, which assigns a meaning to a sequence of elements of the pattern notation:

*A sequence of elements of the pattern notation matches a sequence of amino acids if and only if the latter sequence can be partitioned into subsequences in such a way that each pattern element matches the corresponding subsequence in turn.*

Thus the pattern [AB] [CDE] F matches the six amino acid sequences corresponding to ACF, ADF, AEF, BCF, BDF, and BEF.

Different pattern description notations have other ways of forming pattern elements. One of these notations is the PROSITE notation, described in the following subsection.

### *Prosite pattern notation*

The PROSITE notation uses the IUPAC one-letter codes and conforms to the above description with the exception that a concatenation symbol, '-', is used between pattern elements, but it is often dropped between letters of the pattern alphabet.

Prosite allows the following pattern elements in addition to those described previously:

- The lower case letter 'x' can be used as a pattern element to denote any amino acid.
- A string of characters drawn from the alphabet and enclosed in braces (curly brackets) denotes any amino acid except for those in the string. For example, {ST} denotes any amino acid other than S or T.
- If a pattern is restricted to the N-terminal of a sequence, the pattern is prefixed with '<'.
- If a pattern is restricted to the C-terminal of a sequence, the pattern is suffixed with '>'.
- The character '>' can also occur inside a terminating square bracket pattern, so that S[T>] matches both "ST" and "S>".
- If e is a pattern element, and m and n are two decimal integers with m <= n, then:
  - o e(m) is equivalent to the repetition of e exactly m times;
  - o e(m,n) is equivalent to the repetition of e exactly k times for any integer k satisfying: m <= k <= n.

Some examples:

- x(3) is equivalent to x-x-x.
- x(2,4) matches any sequence that matches x-x or x-x-x or x-x-x-x.

The signature of the C2H2-type *zinc finger* domain is:

- C-x(2,4)-C-x(3)-[LIVMFYWC]-x(8)-H-x(3,5)-H

### *Matrices*

A matrix of numbers containing scores for each residue or nucleotide at each position of a fixed-length motif. There are two types of weight matrices.

- A position frequency matrix (PFM) records the position-dependent frequency of each residue or nucleotide. PFMs can be experimentally determined from SELEX experiments or computationally discovered by tools such as MEME using hidden Markov models.

- A position weight matrix (PWM) contains log odds weights for computing a match score. A cutoff is needed to specify whether an input sequence matches the motif or not. PWMs are calculated from PFMs.

An example of a PFM from the TRANSFAC database for the transcription factor AP-1:

| *Pos* | *A* | *C* | *G* | *T* | *IUPAC* |
|---|---|---|---|---|---|
| 01 | 6 | 2 | 8 | 1 | R |
| 02 | 3 | 5 | 9 | 0 | S |
| 03 | 0 | 0 | 0 | 17 | T |
| 04 | 0 | 0 | 17 | 0 | G |
| 05 | 17 | 0 | 0 | 0 | A |
| 06 | 0 | 16 | 0 | 1 | C |
| 07 | 3 | 2 | 3 | 9 | T |
| 08 | 4 | 7 | 2 | 4 | N |
| 09 | 9 | 6 | 1 | 1 | M |
| 10 | 4 | 3 | 7 | 3 | N |
| 11 | 6 | 3 | 1 | 7 | W |

The first column specifies the position, the second column contains the number of occurrences of A at that position, the third column contains the number of occurrences of C at that position, the fourth column contains the number of occurrences of G at that position, the fifth column contains the number of occurrences of T at that position, and the last column contains the IUPAC notation for that position. Note that the sums of occurrences for A, C, G, and T for each row should be equal because the PFM is derived from aggregating several consensus sequences.

### Another Scheme

The following example comes from the paper by Matsuda, et al. 1997.

The *E. coli* lactose operon repressor LacI (PDB 1lcc chain A) and *E. coli* catabolite gene activator (PDB 3gap chain A) both have a *helix-turn-helix* motif, but their amino acid sequences do not show much similarity, as shown in the table below. Matsuda, et al. devised a code they called the "three-dimensional chain code" for representing a protein structure as a string of letters. This encoding scheme reveals the similarity between the proteins much more clearly than the amino acid sequence:

| | *3D chain code* | *Amino acid sequence* |
|---|---|---|
| 1lccA | TWWWWWWWKCLKWWWWWWG | LYDVAEYAGVSYQTVSRVV |
| 3gapA | KWWWWWWGKCFKWWWWWWW | RQEIGQIVGCSRETVGRIL |

where "W" corresponds to an α-helix, and "E" and "D" correspond to a β-strand.

## Genome Annotation

In the context of genomics, annotation is the process of marking the genes and other biological features in a DNA sequence. The first genome annotation software system was designed in 1995 by Dr. Owen White, who was part of the team at The Institute for Genomic Research that sequenced and analysed the first genome of a free-living organism to be decoded, the bacterium *Haemophilus influenzae*.

Dr. White built a software system to find the genes (places in the DNA sequence that encode a protein), the transfer RNA, and other features, and to make initial assignments of function to those genes. Most current genome annotation systems work similarly, but the programs available for analysis of genomic DNA are constantly changing and improving.

## Gene Finding

Gene finding typically refers to the area of computational biology that is concerned with algorithmically identifying stretches of sequence, usually genomic DNA, that are biologically functional. This especially includes protein-coding genes, but may also include other functional elements such as RNA genes and regulatory regions. Gene finding is one of the first and most important steps in understanding the genome of a species once it has been sequenced.

In its earliest days, "gene finding" was based on painstaking experimentation on living cells and organisms. Statistical analysis of the rates of homologous recombination of several different genes could determine their order on a certain chromosome, and information from many such experiments could be combined to create a genetic map specifying the rough location of known genes relative to each other. Today, with comprehensive genome sequence and powerful computational resources at the disposal of the research community, gene finding has been redefined as a largely computational problem.

Determining that a sequence *is functional* should be distinguished from determining *the function* of the gene or its product. The latter still demands *in vivo* experimentation through gene knockout and other assays, although frontiers of bioinformatics research are making it increasingly possible to predict the function of a gene based on its sequence alone.

### *Extrinsic Approaches*

In extrinsic (or evidence-based) gene finding systems, the target genome is searched for sequences that are similar to extrinsic evidence in the form of the known sequence of a messenger RNA (mRNA) or protein product. Given an mRNA sequence, it is trivial to derive a unique genomic DNA sequence from which it had to have been transcribed. Given a protein sequence, a family of possible coding DNA sequences can be derived by reverse translation of the genetic code. Once candidate DNA sequences have been determined, it is a relatively straightforward algorithmic problem to efficiently search a target genome for matches, complete or partial, and exact or inexact. BLAST is a widely used system designed for this purpose.

A high degree of similarity to a known messenger RNA or protein product is strong evidence that a region of a target genome is a protein-coding gene. However, to apply this approach systemically requires extensive sequencing of mRNA and protein products. Not only is this expensive, but in complex organisms, only a subset of all genes in the organism's genome are expressed at any given time, meaning that extrinsic evidence for many genes is not readily accessible in any single cell culture. Thus, in order to collect extrinsic evidence for most or all of the genes in a complex organism, many hundreds or thousands of different cell types must be studied, which itself presents further difficulties. For example, some human genes may be expressed only during development as an embryo or fetus, which might be difficult to study for ethical reasons.

Despite these difficulties, extensive transcript and protein sequence databases have been generated for human as well as other important model organisms in biology, such as mice and yeast. For example, the RefSeq database contains transcript and protein sequence from many different species, and the Ensembl system comprehensively maps this evidence to human and several other genomes. It is, however, likely that these databases are both incomplete and contain small but significant amounts of erroneous data..

## Ab Initio *Approaches*

Because of the inherent expense and difficulty in obtaining extrinsic evidence for many genes, it is also necessary to resort to *Ab initio* gene finding, in which genomic DNA sequence alone is systematically searched for certain tell-tale signs of protein-coding genes. These signs can be broadly categorized as either *signals*, specific sequences that indicate the presence of a gene nearby, or *content*, statistical properties of protein-coding sequence itself. *Ab initio* gene finding might be more accurately characterized as gene *prediction*, since extrinsic evidence is generally required to conclusively establish that a putative gene is functional. In the genomes of prokaryotes, genes have specific and relatively well-understood promoter sequences (signals), such as the Pribnow box and transcription factor binding sites, which are easy to systematically identify.

Also, the sequence coding for a protein occurs as one contiguous open reading frame (ORF), which is typically many hundred or thousands of base pairs long. The statistics of stop codons are such that even finding an open reading frame of this length is a fairly informative sign. (Since 3 of the 64 possible codons in the genetic code are stop codons, one would expect a stop codon approximately every 20-25 codons, or 60-75 base pairs, in a random sequence.) Furthermore, protein-coding DNA has certain periodicities and other statistical properties that are easy to detect in sequence of this length. These characteristics make prokaryotic gene finding relatively straightforward, and well-designed systems are able to achieve high levels of accuracy.

*Ab initio* gene finding in eukaryotes, especially complex organisms like humans, is considerably more challenging for several reasons. First, the promoter and other regulatory signals in these genomes are more complex and less well-understood than in prokaryotes, making them more difficult to reliably recognize. Two classic examples of signals identified by eukaryotic gene finders are CpG islands and binding sites for a poly(A) tail.

Second, splicing mechanisms employed by eukaryotic cells mean that a particular protein-coding sequence in the genome is divided into several parts (exons), separated by non-coding sequences (introns). (Splice sites are themselves another signal that eukaryotic gene finders are often designed to identify.) A typical protein-coding gene in humans might be divided into a dozen exons, each less than two hundred base pairs in length, and some as short as twenty to thirty. It is therefore much more difficult to detect periodicities and other known content properties of protein-coding DNA in eukaryotes.

Advanced gene finders for both prokaryotic and eukaryotic genomes typically use complex probabilistic models, such as hidden Markov models, in order to combine information from a variety of different signal and content measurements. The GLIMMER system is a widely used and highly accurate gene finder for prokaryotes. GeneMark is another popular approach. Eukaryotic *ab initio* gene finders, by comparison, have achieved only limited success; notable examples are the GENSCAN and geneid programs.

The SNAP gene finder is HMM-based like Genscan and attempts to be more adaptable to different organisms, addressing problems related to using a gene finder on a genome sequence that it was not trained against. A few recent approaches like mSplicer, CONTRAST, or mGene also use machine learning techniques like support vector machines for successful gene prediction. They build a discriminative model using hidden Markov support vector machines or conditional random fields to learn an accurate gene prediction scoring function.

### Other Signals

Among the derived signals used for prediction are statistics resulting from the sub-sequence statistics like k-mer statistics, Fourier transform of a pseudo-number-coded DNA, Z-curve parameters and certain run features.

It has been suggested that signals other than those directly detectable in sequences may improve gene prediction. For example, the role of secondary structure in the identification of regulatory motifs has been reported. In addition, it has been suggested that RNA secondary structure prediction helps splice site prediction.

### Comparative Genomics Approaches

As the entire genomes of many different species are sequenced, a promising direction in current research on gene finding is a comparative genomics approach. This is based on the principle that the forces of natural selection cause genes and other functional elements to undergo mutation at a slower rate than the rest of the genome, since mutations in functional elements are more likely to negatively impact the organism than mutations elsewhere. Genes can thus be detected by comparing the genomes of related species to detect this evolutionary pressure for conservation. This approach was first applied to the mouse and human genomes, using programs such as SLAM, SGP and Twinscan/N-SCAN.

Comparative gene finding can also be used to project high quality annotations from one genome to another. Notable examples include

Projector, GeneWise and GeneMapper. Such techniques now play a central role in the annotation of all genomes.

### *Computational Evolutionary Biology*

Evolutionary biology is the study of the origin and descent of species, as well as their change over time. Informatics has assisted evolutionary biologists in several key ways; it has enabled researchers to:

- trace the evolution of a large number of organisms by measuring changes in their DNA, rather than through physical taxonomy or physiological observations alone,
- more recently, compare entire genomes, which permits the study of more complex evolutionary events, such as gene duplication, horizontal gene transfer, and the prediction of factors important in bacterial speciation,
- build complex computational models of populations to predict the outcome of the system over time
- track and share information on an increasingly large number of species and organisms.

Future work endeavours to reconstruct the now more complex tree of life. The area of research within computer science that uses genetic algorithms is sometimes confused with computational evolutionary biology, but the two areas are not necessarily related.

### *Analysis of Gene Expression*

The expression of many genes can be determined by measuring mRNA levels with multiple techniques including microarrays, expressed cDNA sequence tag (EST) sequencing, serial analysis of gene expression (SAGE) tag sequencing, massively parallel signature sequencing (MPSS), or various applications of multiplexed in-situ hybridization. All of these techniques are extremely noise-prone and/or subject to bias in the biological measurement, and a major research area in computational biology involves developing statistical tools to separate signal from noise in high-throughput gene expression studies. Such studies are often used to determine the genes implicated in a disorder: one might compare microarray data from cancerous epithelial cells to data from non-cancerous cells to determine the transcripts that are up-regulated and down-regulated in a particular population of cancer cells.

### *Analysis of Regulation*

Regulation is the complex orchestration of events starting with an extracellular signal such as a hormone and leading to an increase or decrease in the activity of one or more proteins. Bioinformatics techniques have been applied to explore various steps in this process. For example, promoter analysis involves the identification and study of sequence motifs in the DNA surrounding the coding region of a gene. These motifs influence the extent to which that region is transcribed into mRNA. Expression data can be used to infer gene regulation: one might compare microarray data from a wide variety of states of an organism to form hypotheses about the genes involved in each state. In a single-cell organism, one might compare stages of the cell cycle, along with various stress conditions (heat shock, starvation, etc.). One can then apply clustering algorithms to that expression data to determine which genes are co-expressed. For example, the upstream regions (promoters) of co-expressed genes can be searched for over-represented regulatory elements.

### Analysis of Protein Expression

Protein microarrays and high throughput (HT) mass spectrometry (MS) can provide a snapshot of the proteins present in a biological sample. Bioinformatics is very much involved in making sense of protein microarray and HT MS data; the former approach faces similar problems as with microarrays targeted at mRNA, the latter involves the problem of matching large amounts of mass data against predicted masses from protein sequence databases, and the complicated statistical analysis of samples where multiple, but incomplete peptides from each protein are detected.

### Analysis of Mutations in Cancer

In cancer, the genomes of affected cells are rearranged in complex or even unpredictable ways. Massive sequencing efforts are used to identify previously unknown point mutations in a variety of genes in cancer. Bioinformaticians continue to produce specialized automated systems to manage the sheer volume of sequence data produced, and they create new algorithms and software to compare the sequencing results to the growing collection of human genome sequences and germline polymorphisms. New physical detection technologies are employed, such as oligonucleotide microarrays to identify chromosomal gains and losses (called comparative genomic hybridization), and single-nucleotide polymorphism arrays to detect known *point mutations*. These detection methods simultaneously measure several hundred thousand sites throughout the genome, and when used in high-throughput to measure thousands of samples, generate terabytes of data per experiment. Again the massive amounts and new types of data generate new opportunities for bioinformaticians. The data is often found to contain considerable variability, or noise, and thus Hidden Markov model and change-point analysis methods are being developed to infer real copy number changes. Another type of data that requires novel informatics development is the analysis of lesions found to be recurrent among many tumours.

### *Comparative Genomics*

The core of comparative genome analysis is the establishment of the correspondence between genes (orthology analysis) or other genomic features in different organisms. It is these intergenomic maps that make it possible to trace the evolutionary processes responsible for the divergence of two genomes. A multitude of evolutionary events acting at various organizational levels shape genome evolution.

At the lowest level, point mutations affect individual nucleotides. At a higher level, large chromosomal segments undergo duplication, lateral transfer, inversion, transposition, deletion and insertion. Ultimately, whole genomes are involved in processes of hybridization, polyploidization and endosymbiosis, often leading to rapid speciation. The complexity of genome evolution poses many exciting challenges to developers of mathematical models and algorithms, who have recourse to a spectra of algorithmic, statistical and mathematical techniques, ranging from exact, heuristics, fixed parameter and approximation algorithms for problems based on parsimony models to Markov Chain Monte Carlo algorithms for Bayesian analysis of problems based on probabilistic models. Many of these studies are based on the homology detection and protein families computation.

### *Modelling Biological Systems*

Systems biology involves the use of computer simulations of cellular subsystems (such as the networks of metabolites and enzymes which comprise metabolism, signal transduction pathways and gene regulatory networks) to both analyse and visualize the complex connections of these cellular processes. Artificial life or virtual evolution attempts to understand evolutionary processes via the computer simulation of simple (artificial) life forms.

## Computational Systems Biology

Modelling biological systems is a significant task of systems biology and mathematical biology. Computational systems biology aims to develop and use efficient algorithms, data structures, visualization and communication tools to orchestrate the integration of large quantities of biological data with the goal of computer modelling. It involves the use of computer simulations of biological systems, like cellular subsystems (such as the networks of metabolites and enzymes which comprise metabolism, signal transduction pathways and gene regulatory networks) to both analyse and visualize the complex connections of these cellular processes. It is also directly associated with bioinformatics and computational biology. Artificial life or virtual evolution attempts to understand evolutionary processes via the computer simulation of simple (artificial) life forms.

It is understood that an unexpected emergent property of a complex system is a result of the interplay of the cause-and-effect among simpler, integrated parts. Biological systems manifest many important examples of emergent properties in the complex interplay of components. Traditional study of biological systems requires reductive methods in which quantities of data are gathered by category, such as concentration over time in response to a certain stimulus. Computers are critical to analysis and modelling of these data. The goal is to create accurate real-time models of a system's response to environmental and internal stimuli, such as a model of a cancer cell in order to find weaknesses in its signalling pathways, or modelling of ion channel mutations to see effects on cardiomyocytes and in turn, the function of a beating heart. A monograph on this topic summarizes an extensive amount of published research in this area up to 1987, including subsections in the following areas: computer modelling in biology and medicine, arterial system models, neuron models, biochemical and oscillation networks, quantum automata, quantum computers in molecular biology and genetics, cancer modelling, neural nets, genetic networks, abstract relational biology, metabolic-replication systems, category theory applications in biology and medicine, automata theory, cellular automata, tessallation models and complete self-reproduction, chaotic systems in organisms, relational biology and organismic theories. This published report also includes 390 references to peer-reviewed articles by a large number of authors.

### *Standards*

By far the most widely accepted standard format for storing and exchanging models in the field is the Systems Biology Markup Language (SBML) The SBML.org website includes a guide to many important software packages used in computational systems biology. Other markup languages with different emphases include BioPAX and CellML.

## Structural Bioinformatic Approaches

### *Prediction of Protein Structure*

Protein structure prediction is another important application of bioinformatics. The amino acid sequence of a protein, the so-called primary structure, can be easily determined from the sequence on the gene that codes for it. In the vast majority of cases, this primary structure uniquely determines a structure in its native environment. (Of course, there are exceptions, such as the bovine spongiform encephalopathy-aka Mad Cow Disease-prion.) Knowledge of this structure is vital in understanding the function of the protein. For lack of better terms, structural information is usually classified as one of *secondary*, *tertiary* and *quaternary* structure. A viable general solution to such predictions remains an open problem. As of now, most efforts have been directed towards heuristics that work most of the time.

One of the key ideas in bioinformatics is the notion of homology. In the genomic branch of bioinformatics, homology is used to predict the function of a gene: if the sequence of gene *A*, whose function is known, is homologous to the sequence of gene *B*, whose function is unknown, one could infer that B may share A's function. In the structural branch of bioinformatics, homology is used to determine which parts of a protein are important in structure formation and interaction with other proteins.

In a technique called homology modelling, this information is used to predict the structure of a protein once the structure of a homologous protein is known. This currently remains the only way to predict protein structures reliably. One example of this is the similar protein homology between hemoglobin in humans and the hemoglobin in legumes (leghemoglobin). Both serve the same purpose of transporting oxygen in the organism. Though both of these proteins have completely different amino acid sequences, their protein structures are virtually identical, which reflects their near identical purposes.

Other techniques for predicting protein structure include protein threading and *de novo* (from scratch) physics-based modelling.

## Structural Domain

A protein domain is a part of protein sequence and structure that can evolve, function, and exist independently of the rest of the protein chain. Each domain forms a compact three-dimensional structure and often can be independently stable and folded. Many proteins consist of several structural domains. One domain may appear in a variety of different proteins.

Molecular evolution uses domains as building blocks and these may be recombined in different arrangements to create proteins with different functions. Domains vary in length from between about 25 amino acids up to 500 amino acids in length. The shortest domains such as zinc fingers are stabilized by metal ions or disulfide bridges. Domains often form functional units, such as the calcium-binding EF hand domain of calmodulin. Because they are independently stable, domains can be "swapped" by genetic engineering between one protein and another to make chimeric proteins.

The concept of the domain was first proposed in 1973 by Wetlaufer after X-ray crystallographic studies of hen lysozyme and papain and by limited proteolysis studies of immunoglobulins. Wetlaufer defined domains as stable units of protein structure that could fold autonomously. In the past domains have been described as units of:

- compact structure
- function and evolution
- folding.

Each definition is valid and will often overlap, i.e. a compact structural domain that is found amongst diverse proteins is likely to fold independently within its structural environment. Nature often brings several domains together to form multidomain and multifunctional proteins with a vast number of possibilities.

In a multidomain protein, each domain may fulfil its own function independently, or in a concerted manner with its neighbours. Domains can either serve as modules for building up large assemblies such as virus particles or muscle fibres, or can provide specific catalytic or binding sites as found in enzymes or regulatory proteins.

An appropriate example is pyruvate kinase, a glycolytic enzyme that plays an important role in regulating the flux from fructose-1,6-biphosphate to pyruvate. It contains an all-β regulatory domain, an α/β-substrate binding domain and an α/β-nucleotide binding domain, connected by several polypeptide linkers. Each domain in this protein occurs in diverse sets of protein families.

The central α/β-barrel substrate binding domain is one of the most common enzyme folds. It is seen in many different enzyme families catalysing completely unrelated reactions. The α/β-barrel is commonly called the TIM barrel named after triose phosphate isomerase, which was the first such structure to be solved. It is currently classified into 26 homologous families in the CATH domain database. The TIM barrel is formed from a sequence of β-α-β motifs closed by the first and last strand hydrogen bonding together, forming an eight stranded barrel. There is debate about the evolutionary origin of this domain. One study has suggested that a single ancestral enzyme could have diverged into several families, while another suggests that a stable TIM-barrel structure has evolved through convergent evolution.

The TIM-barrel in pyruvate kinase is 'discontinuous', meaning that more than one segment of the polypeptide is required to form the domain. This is likely to be the result of the insertion of one domain into another during the protein's evolution. It has been shown from known structures that about a quarter of structural domains are discontinuous. The inserted β-barrel regulatory domain is 'continuous', made up of a single stretch of polypeptide.

Covalent association of two domains represents a functional and structural advantage since there is an increase in stability when compared with the same structures non-covalently associated. Other, advantages are the protection of intermediates within inter-domain enzymatic clefts that may otherwise be unstable in aqueous environments, and a fixed stoichiometric ratio of the enzymatic activity necessary for a sequential set of reactions.

### *Domains are Units of Protein Structure*

The primary structure (string of amino acids) of a protein ultimately encodes its uniquely folded 3D conformation. The most important factor governing the folding of a protein into 3D structure is the distribution of polar and non-polar side chains. Folding is driven by the burial of hydrophobic side chains into the interior of the molecule so to avoid contact with the aqueous environment. Generally proteins have a core of hydrophobic residues surrounded by a shell of hydrophilic residues. Since the peptide bonds themselves are polar they are neutralised by hydrogen bonding with each other when in the hydrophobic environment. This gives rise to regions of the polypeptide that form regular 3D structural patterns called secondary structure. There are two main types of secondary structure: α-helices and β-sheets.

Some simple combinations of secondary structure elements have been found to frequently occur in protein structure and are referred to as supersecondary structure or motifs. For example, the β-hairpin motif consists of two adjacent antiparallel β-strands joined by a small loop. It is present in most antiparallel β structures both as an isolated ribbon and as part of more complex β-sheets. Another common super-secondary structure is the β-α-β motif, which is frequently used to connect two parallel β-strands. The central α-helix connects the C-termini of the first strand to the N-termini of the second strand, packing its side chains against the β-sheet and therefore shielding the hydrophobic residues of the β-strands from the surface.

Structural alignment is an important tool for determining domains.

### Tertiary Structure of Domains

Several motifs pack together to form compact, local, semi-independent units called domains. The overall 3D structure of the polypeptide chain is referred to as the protein's tertiary structure. Domains are the fundamental units of tertiary structure, each domain containing an individual hydrophobic core built from secondary structural units connected by loop regions. The packing of the polypeptide is usually much tighter in the interior than the exterior of the domain producing a solid-like core and a fluid-like surface. In fact, core residues are often conserved in a protein family, whereas the residues in loops are less conserved, unless they are involved in the protein's function. Protein tertiary structure can be divided into four main classes based on the secondary structural content of the domain.

- All-α domains have a domain core built exclusively from α-helices. This class is dominated by small folds, many of which form a simple bundle with helices running up and down.
- All-β domains have a core comprising of antiparallel β-sheets, usually two sheets packed against each other. Various patterns can be identified in the arrangement of the strands, often giving rise to the identification of recurring motifs, for example the Greek key motif.
- α+β domains are a mixture of all-α and all-β motifs. Classification of proteins into this class is difficult because of overlaps to the other three classes and therefore is not used in the CATH domain database.
- α/β domains are made from a combination of β-α-β motifs that predominantly form a parallel β-sheet surrounded by amphipathic α-helices. The secondary structures are arranged in layers or barrels.

### Domains have Limits on Size

Domains have limits on size. The size of individual structural domains varies from 36 residues in E-selectin to 692 residues in lipoxygenase-1, but the majority, 90%, have less than 200 residues with an average of approximately 100 residues. Very short domains, less than 40 residues, are often stabilised by metal ions or disulfide bonds. Larger domains, greater than 300 residues, are likely to consist of multiple hydrophobic cores.

### Domains and Quaternary Structure

Many proteins have a quaternary structure, which consists of several polypeptide chains that associate into an oligomeric molecule.

Each polypeptide chain in such a protein is called a subunit. Hemoglobin, for example, consists of two α and two β subunits. Each of the four chains has an all-α globin fold with a heme pocket.

Domain swapping is a mechanism for forming oligomeric assemblies. In domain swapping, a secondary or tertiary element of a monomeric protein is replaced by the same element of another protein. Domain swapping can range from secondary structure elements to whole structural domains. It also represents a model of evolution for functional adaptation by oligomerisation, e.g. oligomeric enzymes that have their active site at subunit interfaces.

### Domains as Evolutionary Modules

*Nature is a tinkerer and not an inventor*, new sequences are adapted from pre-existing sequences rather than invented. Domains are the common material used by nature to generate new sequences, they can be thought of as genetically mobile units, referred to as 'modules'. Often, the C and N termini of domains are close together in space, allowing them to easily be "slotted into" parent structures during the process of evolution. Many domain families are found in all three forms of life, Archaea, Bacteria and Eukarya. Domains that are repeatedly found in diverse proteins are often referred to as modules, examples can be found among extracellular proteins associated with clotting, fibrinolysis, complement, the extracellular matrix, cell surface adhesion molecules and cytokine receptors.

Molecular evolution gives rise to families of related proteins with similar sequence and structure. However, sequence similarities can be extremely low between proteins that share the same structure. Protein structures may be similar because proteins have diverged from a common ancestor.

Alternatively, some folds may be more favoured than others as they represent stable arrangements of secondary structures and some proteins may converge towards these folds over the course of evolution. There are currently about 45,000 experimentally determined protein 3D structures deposited within the Protein Data Bank (PDB). However this set contains a lot of identical or very similar structures. All proteins should be classified to structural families to understand their evolutionary relationships. Structural comparisons are best achieved at the domain level. For this reason many algorithms have been developed to automatically assign domains in proteins with known 3D structure. The CATH domain database classifies domains into approximately 800 fold families, ten of these folds are highly populated and are referred to as 'super-folds'. Super-folds are defined as folds for which there are at least three structures without significant sequence similarity. The most populated is the α/β-barrel super-fold as described previously.

### *Multidomain Proteins*

The majority of genomic proteins, two-thirds in unicellular organisms and more than 80% in metazoa, are multidomain proteins created as a result of gene duplication events. Many domains in multidomain structures could have once existed as independent proteins. More and more domains in eukaryotic multidomain proteins can be found as independent proteins in prokaryotes.

For example, vertebrates have a multi-enzyme polypeptide containing the GAR synthetase, AIR synthetase and GAR transformylase modules (GARs-AIRs-GARt; GAR: glycinamide ribonucleotide synthetase/ transferase; AIR: aminoimidazole ribonucleotide synthetase). In insects, the polypeptide appears as GARs-(AIRs)2-GARt, in yeast GARs-AIRs is encoded separately from GARt, and in bacteria each domain is encoded separately.

### *Origin*

Multidomain proteins are likely to have emerged from a selective pressure during evolution to create new functions. Various proteins have diverged from common ancestors by different combinations and associations of domains. Modular units frequently move about, within and between biological systems through mechanisms of genetic shuffling:

- transposition of mobile elements including horizontal transfers (between species);
- gross rearrangements such as inversions, translocations, deletions and duplications;
- homologous recombination;
- slippage of DNA polymerase during replication.

### *Types of Organisation*

The simplest multidomain organisation seen in proteins is that of a single domain repeated in tandem. The domains may interact with each other or remain isolated, like beads on string. The giant 30,000 residue muscle protein titin comprises about 120 fibronectin-III-type and Ig-type domains. In the serine proteases, a gene duplication event has led to the formation of a two β-barrel domain enzyme. The repeats have diverged so widely that there is no obvious sequence similarity between them. The active site is located at a cleft between the two β-barrel domains, in which functionally important residues are contributed from each domain. Genetically engineered mutants of the chymotrypsin serine protease were shown to have some proteinase activity even though their active site residues were abolished and it has therefore been postulated that the duplication event enhanced the enzyme's activity.

Modules frequently display different connectivity relationships, as illustrated by the kinesins and ABC transporters. The kinesin motor domain can be at either end of a polypeptide chain that includes a coiled-coil region and a cargo domain. ABC transporters are built with up to four domains consisting of two unrelated modules, ATP-binding cassette and an integral membrane module, arranged in various combinations. Not only do domains recombine, but there are many examples of a domain having been inserted into another.

Sequence or structural similarities to other domains demonstrate that homologues of inserted and parent domains can exist independently. An example is that of the 'fingers' inserted into the 'palm' domain within the polymerases of the Pol I family. Since a domain can be inserted into another, there should always be at least one continuous domain in a multidomain protein. This is the main difference between definitions of structural domains and evolutionary/functional domains. An evolutionary domain will be limited to one or two connections between domains, whereas structural domains can have unlimited connections, within a given criterion of the existence of a common core. Several structural domains could be assigned to an evolutionary domain.

## Domains are Autonomous Folding Units

### *Folding*

Protein folding-the unsolved problem : Since the seminal work of Anfinsen over forty years ago, the goal to completely understand the mechanism by which a polypeptide rapidly folds into its stable native conformation remains elusive. Many experimental folding studies have contributed much to our understanding, but the principles that govern protein folding are still based on those discovered in the very first studies of folding. Anfinsen showed that the native state of a protein is thermodynamically stable, the conformation being at a global minimum of its free energy.

Folding is a directed search of conformational space allowing the protein to fold on a biologically feasible time scale. The Levinthal paradox states that if an averaged sized protein would sample all possible conformations before finding the one with the lowest energy, the whole process would take billions of years. Proteins typically fold within 0.1 and 1000 seconds, therefore the protein folding process must be directed some way through a specific folding pathway. The forces that direct this search are likely to be a combination of local and global influences whose effects are felt at various stages of the reaction.

Advances in experimental and theoretical studies have shown that folding can be viewed in terms of energy landscapes, where folding kinetics is considered as a progressive organisation of an ensemble of partially folded structures through which a protein passes on its way to the folded structure.

This has been described in terms of a folding funnel, in which an unfolded protein has a large number of conformational states available and there are fewer states available to the folded protein. A funnel implies that for protein folding there is a decrease in energy and loss of entropy with increasing tertiary structure formation. The local roughness of the funnel reflects kinetic traps, corresponding to the accumulation of misfolded intermediates. A folding chain progresses toward lower intra-chain free-energies by increasing its compactness. The chains conformational options become increasingly narrowed ultimately toward one native structure.

### *Advantage of Domains in Protein Folding*

The organisation of large proteins by structural domains represents an advantage for protein folding, with each domain being able to individually fold, accelerating the folding process and reducing a potentially large combination of residue interactions. Furthermore, given the observed random distribution of hydrophobic residues in proteins, domain formation appears to be the optimal solution for a large protein to bury its hydrophobic residues while keeping the hydrophilic residues at the surface.

However, the role of inter-domain interactions in protein folding and in energetics of stabilisation of the native structure, probably differs for each protein. In T4 lysozyme, the influence of one domain on the other is so strong that the entire molecule is resistant to proteolytic cleavage. In this case, folding is a sequential process where the C-terminal domain is required to fold independently in an early step, and the other domain requires the presence of the folded C-terminal domain for folding and stabilisation. It has been found that the folding of an isolated domain can take place at the same rate or sometimes faster than that of the integrated domain. Suggesting that unfavourable interactions with the rest of the protein can occur during folding.

Several arguments suggest that the slowest step in the folding of large proteins is the pairing of the folded domains. This is either because the domains are not folded entirely correctly or because the small adjustments required for their interaction are energetically unfavourable, such as the removal of water from the domain interface.

## Domains and Protein Flexibility

The presence of multiple domains in proteins gives rise to a great deal of flexibility and mobility, leading to protein domain dynamics. Domain motions can be inferred by comparing different structures of a protein, or they can be directly observed using spectra measured by neutron spin echo spectroscopy. They can also be suggested by sampling in extensive molecular dynamics trajectories. Domain motions are important for:

- catalysis;
- regulatory activity;
- transport of metabolites;
- formation of protein assemblies; and
- cellular locomotion.

One of the largest observed domain motions is the 'swivelling' mechanism in pyruvate phosphate dikinase. The phosphoinositide domain swivels between two states in order to bring a phosphate group from the active site of the nucleotide binding domain to that of the phosphoenolpyruvate/pyruvate domain. The phosphate group is moved over a distance of 45A involving a domain motion of about 100 degrees around a single residue. In enzymes, the closure of one domain onto another captures a substrate by an induced fit, allowing the reaction to take place in a controlled way. A detailed analysis by Gerstein led to the classification of two basic types of domain motion; hinge and shear. Only a relatively small portion of the chain, namely the inter-domain linker and side chains undergo significant conformational changes upon domain rearrangement.

### *Hinges by Secondary Structures*

A study by Hayward found that the termini of α-helices and β-sheets form hinges in a large number of cases. Many hinges were found to involve two secondary structure elements acting like hinges of a door, allowing an opening and closing motion to occur. This can arise when two neighbouring strands within a β-sheet situated in one domain, diverge apart as they join the other domain. The two resulting termini then form the bending regions between the two domains. α-helices that preserve their hydrogen bonding network when bent are found to behave as mechanical hinges, storing 'elastic energy' that drives the closure of domains for rapid capture of a substrate.

### Helical to Extended Conformation

The interconversion of helical and extended conformations at the site of a domain boundary is not uncommon. In calmodulin, torsion angles change for five residues in the middle of a domain linking α-helix. The helix is split into two, almost perpendicular, smaller helices separated by four residues of an extended strand.

### Shear Motions

Shear motions involve a small sliding movement of domain interfaces, controlled by the amino acid side chains within the interface. Proteins displaying shear motions often have a layered architecture: stacking of secondary structures. The interdomain linker has merely the role of keeping the domains in close proximity.

### Domain Motion and Functional Dynamics in Enzymes

The analysis of the internal dynamics of structurally different, but functionally similar enzymes has highlighted a common relationship between the positioning of the active site and the two principal protein sub-domains. In fact, for several members of the hydrolase superfamily, the catalytic site is located close to the interface separating the two principal quasi-rigid domains. Such positioning appears instrumental for maintaining the precise geometry of the active site, while allowing for an appreciable functionally-oriented modulation of the flanking regions resulting from the relative motion of the two sub-domains.

### Domain Definition from Structural Co-ordinates

The importance of domains as structural building blocks and elements of evolution has brought about many automated methods for their identification and classification in proteins of known structure. Automatic procedures for reliable domain assignment is essential for the generation of the domain databases, especially as the number of protein structures is increasing.

Although the boundaries of a domain can be determined by visual inspection, construction of an automated method is not straightforward. Problems occur when faced with domains that are discontinuous or highly associated. The fact that there is no standard definition of what a domain really is has meant that domain assignments have varied enormously, with each researcher using a unique set of criteria.

A structural domain is a compact, globular sub-structure with more interactions within it than with the rest of the protein. Therefore, a structural domain can be determined by two visual characteristics; its compactness and its extent of isolation. Measures of local compactness in proteins have been used in many of the early methods of domain assignment and in several of the more recent methods.

### *Methods*

One of the first algorithms used a Cα-Cα distance map together with a hierarchical clustering routine that considered proteins as several small segments, 10 residues in length. The initial segments were clustered one after another based on inter-segment distances; segments with the shortest distances were clustered and considered as single segments thereafter. The stepwise clustering finally included the full protein. Go also exploited the fact that inter-domain distances are normally larger than intra-domain distances; all possible Cα-Cα distances were represented as diagonal plots in which there were distinct patterns for helices, extended strands and combinations of secondary structures.

The method by Sowdhamini and Blundell clusters secondary structures in a protein based on their Cα-Cα distances and identifies domains from the pattern in their dendrograms. As the procedure does not consider the protein as a continuous chain of amino acids there are no problems in treating discontinuous domains. Specific nodes in these dendrograms are identified as tertiary structural clusters of the protein, these include both super-secondary structures and domains. The DOMAK algorithm is used to create the 3Dee domain database. It calculates a 'split value' from the number of each type of contact when the protein is divided arbitrarily into two parts. This split value is large when the two parts of the structure are distinct. The method of Wodak and Janin was based on the calculated interface areas between two chain segments repeatedly cleaved at various residue positions. Interface areas were calculated by comparing surface areas of the cleaved segments with that of the native structure. Potential domain boundaries can be identified at a site where the interface area was at a minimum. Other methods have used measures of solvent accessibility to calculate compactness.

The PUU algorithm incorporates a harmonic model used to approximate inter-domain dynamics. The underlying physical concept is that many rigid interactions will occur within each domain and loose interactions will occur between domains. This algorithm is used to define domains in the FSSP domain database.

Swindells (1995) developed a method, DETECTIVE, for identification of domains in protein structures based on the idea that domains have a hydrophobic interior. Deficiencies were found to occur when hydrophobic cores from different domains continue through the interface region. RigidFinder is a novel method for identification of protein rigid blocks (domains and loops) from two different conformations. Rigid blocks are defined as blocks where all inter residue distances are conserved across conformations.

A general method to identify *dynamical domains*, that is protein regions that behave approximately as rigid units in the course of structural fluctuations, has been introduced by Potestio et al. and, among other applications was also used to compare the consistency of the dynamics-based domain subdivisions with standard structure-based ones. The method, termed PiSQRD, is publicly available in the form of a webserver. The latter allows users to optimally subdivide single-chain or multimeric proteins into quasi-rigid domains based on the collective modes of fluctuation of the system. By default the latter are calculated through an elastic network model; alternatively pre-calculated essential dynamical spaces can be uploaded by the user.

### *Molecular Interaction*

Efficient software is available today for studying interactions among proteins, ligands and peptides. Types of interactions most often encountered in the field include-Protein-ligand (including drug), protein-protein and protein-peptide.

Molecular dynamic simulation of movement of atoms about rotatable bonds is the fundamental principle behind computational algorithms, termed docking algorithms for studying molecular interactions.

## Protein-protein Interaction Prediction

### *Docking Algorithms*

In the last two decades, tens of thousands of protein three-dimensional structures have been determined by X-ray crystallography and Protein nuclear magnetic resonance spectroscopy (protein NMR). One central question for the biological scientist is whether it is practical to predict possible protein-protein interactions only based on these 3D shapes, without doing protein-protein interaction experiments. A variety of methods have been developed to tackle the Protein-protein docking problem, though it seems that there is still much work to be done in this field.

### Software and Tools

Software tools for bioinformatics range from simple command-line tools, to more complex graphical programs and standalone web-services available from various bioinformatics companies or public institutions.

### Web Services in Bioinformatics

SOAP and REST-based interfaces have been developed for a wide variety of bioinformatics applications allowing an application running on one computer in one part of the world to use algorithms, data and computing resources on servers in other parts of the world. The main advantages derive from the fact that end users do not have to deal with software and database maintenance overheads.

Basic bioinformatics services are classified by the EBI into three categories: SSS (Sequence Search Services), MSA (Multiple Sequence Alignment) and BSA (Biological Sequence Analysis). The availability of these service-oriented bioinformatics resources demonstrate the applicability of web based bioinformatics solutions, and range from a collection of standalone tools with a common data format under a single, standalone or web-based interface, to integrative, distributed and extensible bioinformatics workflow management systems.

# Chapter 2

# Human Genome Project

The Human Genome Project (HGP) is an international scientific research project with a primary goal of determining the sequence of chemical base pairs which make up DNA and to identify and map the approximately 20,000–25,000 genes of the human genome from both a physical and functional standpoint.

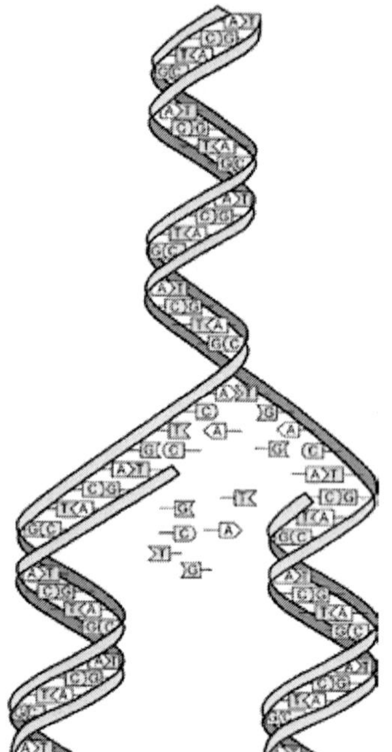

***Figure:** DNA replication*

The project began in 1990 and was initially headed by Ari Patrinos, head of the Office of Biological and Environmental Research in the

U.S. Department of Energy's Office of Science. Francis Collins directed the National Institutes of Health National Human Genome Research Institute efforts. A working draft of the genome was announced in 2000 and a complete one in 2003, with further, more detailed analysis still being published. A parallel project was conducted outside of government by the Celera Corporation, which was formally launched in 1998. Most of the government-sponsored sequencing was performed in universities and research centres from the United States, the United Kingdom, Japan, France, Germany, and China. The mapping of human genes is an important step in the development of medicines and other aspects of health care.

While the objective of the Human Genome Project is to understand the genetic makeup of the human species, the project has also focused on several other nonhuman organisms such as *E. coli*, the fruit fly, and the laboratory mouse. It remains one of the largest single investigative projects in modern science.

The Human Genome Project originally aimed to map the nucleotides contained in a human haploid reference genome (more than three billion). Several groups have announced efforts to extend this to diploid human genomes including the International HapMap Project, Applied Biosystems, Perlegen, Illumina, JCVI, Personal Genome Project, and Roche-454.

The "genome" of any given individual (except for identical twins and cloned organisms) is unique; mapping "the human genome" involves sequencing multiple variations of each gene. The project did not study the entire DNA found in human cells; some heterochromatic areas (about 8% of the total genome) remain un-sequenced.

## Project

The project began with the culmination of several years of work supported by the United States Department of Energy, in particular workshops in 1984 and 1986 and a subsequent initiative of the US Department of Energy. This 1987 report stated boldly, "The ultimate goal of this initiative is to understand the human genome" and "knowledge of the human as necessary to the continuing progress of medicine and other health sciences as knowledge of human anatomy has been for the present state of medicine." Candidate technologies were already being considered for the proposed undertaking at least as early as 1985.

James D. Watson was head of the National Centre for Human Genome Research at the National Institutes of Health (NIH) in the

United States starting from 1988. Largely due to his disagreement with his boss, Bernadine Healy, over the issue of patenting genes, Watson was forced to resign in 1992. He was replaced by Francis Collins in April 1993, and the name of the Centre was changed to the National Human Genome Research Institute (NHGRI) in 1997. The $3-billion project was formally founded in 1990 by the United States Department of Energy and the U.S. National Institutes of Health, and was expected to take 15 years. In addition to the United States, the international consortium comprised geneticists in the United Kingdom, France, Germany, Japan, China, and India.

Due to widespread international cooperation and advances in the field of genomics (especially in sequence analysis), as well as major advances in computing technology, a 'rough draft' of the genome was finished in 2000 (announced jointly by then US president Bill Clinton and the British Prime Minister Tony Blair on June 26, 2000). This first available rough draft assembly of the genome was completed by the UCSC Genome Bioinformatics Group, primarily led by then graduate student Jim Kent. Ongoing sequencing led to the announcement of the essentially complete genome in April 2003, 2 years earlier than planned. In May 2006, another milestone was passed on the way to completion of the project, when the sequence of the last chromosome was published in the journal Nature.

### State of Completion

There are multiple definitions of the "complete sequence of the human genome". According to some of these definitions, the genome has already been completely sequenced, and according to other definitions, the genome has yet to be completely sequenced. There have been multiple popular press articles reporting that the genome was "complete." The genome has been completely sequenced using the definition employed by the International Human Genome Project. A graphical history of the human genome project shows that most of the human genome was complete by the end of 2003. However, there are a number of regions of the human genome that can be considered unfinished:

- First, the central regions of each chromosome, known as centromeres, are highly repetitive DNA sequences that are difficult to sequence using current technology. The centromeres are millions (possibly tens of millions) of base pairs long, and for the most part these are entirely un-sequenced.

- Second, the ends of the chromosomes, called telomeres, are also highly repetitive, and for most of the 46 chromosome ends these too are incomplete. It is not known precisely how much sequence remains before the telomeres of each chromosome are reached, but as with the centromeres, current technological restraints are prohibitive.
- Third, there are several loci in each individual's genome that contain members of multigene families that are difficult to disentangle with shotgun sequencing methods – these multigene families often encode proteins important for immune functions.
- Other than these regions, there remain a few dozen gaps scattered around the genome, some of them rather large, but there is hope that all these will be closed in the next couple of years.

In summary: the best estimates of total genome size indicate that about 92.3% of the genome has been completed and it is likely that the centromeres and telomeres will remain un-sequenced until new technology is developed that facilitates their sequencing. Most of the remaining DNA is highly repetitive and unlikely to contain genes, but it cannot be truly known until it is entirely sequenced. Understanding the functions of all the genes and their regulation is far from complete. The roles of junk DNA, the evolution of the genome, the differences between individuals, and many other questions are still the subject of intense interest by laboratories all over the world.

## Applications and Proposed Benefits

The sequence of the human DNA is stored in databases available to anyone on the Internet. The U.S. National Centre for Biotechnology Information (and sister organizations in Europe and Japan) house the gene sequence in a database known as GenBank, along with sequences of known and hypothetical genes and proteins. Other organizations such as the University of California, Santa Cruz, and Ensembl present additional data and annotation and powerful tools for visualizing and searching it. Computer programs have been developed to analyse the data, because the data itself is difficult to interpret without such programs.

The process of identifying the boundaries between genes and other features in a raw DNA sequence is called genome annotation and is the domain of bioinformatics. While expert biologists make the best annotators, their work proceeds slowly, and computer programs are increasingly used to meet the high-throughput demands of genome sequencing projects. The best current technologies for annotation make use of statistical models that take advantage of parallels between DNA sequences and human language, using concepts from computer science such as formal grammars.

Another, often overlooked, goal of the HGP is the study of its ethical, legal, and social implications. It is important to research these issues and find the most appropriate solutions before they become large dilemmas whose effect will manifest in the form of major political concerns.

All humans have unique gene sequences. Therefore the data published by the HGP does not represent the exact sequence of each and every individual's genome.

It is the combined "reference genome" of a small number of anonymous donors. The HGP genome is a scaffold for future work in identifying differences among individuals. Most of the current effort in identifying differences among individuals involves single-nucleotide polymorphisms and the HapMap.

### *Findings*

Key findings of the draft (2001) and complete (2004) genome sequences include:

1. There are approximately 20,500 genes in human beings, the same range as in mice and twice that of roundworms. Understanding how these genes express themselves will provide clues to how diseases are caused.
2. Between 1.1% to 1.4% of the genome's sequence codes for proteins
3. The human genome has significantly more segmental duplications (nearly identical, repeated sections of DNA) than other mammalian genomes. These sections may underlie the creation of new primate-specific genes
4. At the time when the draft sequence was published less than 7% of protein families appeared to be vertebrate specific.

### *How it was Accomplished*

The first printout of the human genome to be presented as a series of books, displayed at the Wellcome Collection, London.

The Human Genome Project was started in 1989 with the goal of sequencing and identifying all three billion chemical units in the human genetic instruction set, finding the genetic roots of disease and then developing treatments. With the sequence in hand, the next step was to identify the genetic variants that increase the risk for common diseases like cancer and diabetes.

***Figure:*** *The first printout of the human genome to be presented as a series of books, displayed at the Wellcome Collection, London*

It was far too expensive at that time to think of sequencing patients' whole genomes. So the National Institutes of Health embraced the idea for a "shortcut", which was to look just at sites on the genome where many people have a variant DNA unit. The theory behind the shortcut was that since the major diseases are common, so too would be the genetic variants that caused them. Natural selection keeps the human genome free of variants that damage health before children are grown, the theory held, but fails against variants that strike later in life, allowing them to become quite common. (In 2002 the National Institutes of Health started a $138 million project called the HapMap to catalogue the common variants in European, East Asian and African genomes.)

The genome was broken into smaller pieces; approximately 150,000 base pairs in length. These pieces were then ligated into a type of vector known as "bacterial artificial chromosomes", or BACs, which are derived from bacterial chromosomes which have been genetically engineered. The vectors containing the genes can be inserted into bacteria where they are copied by the bacterial DNA replication machinery. Each of these pieces was then sequenced separately as a small "shotgun" project and then assembled. The larger, 150,000 base pairs go together to create chromosomes. This is known as the "hierarchical shotgun" approach, because the genome is first broken into relatively large chunks, which are then mapped to chromosomes before being selected for sequencing.

Funding came from the US government through the National Institutes of Health in the United States, and a UK charity organization, the Wellcome Trust, as well as numerous other groups from around the world. The funding supported a number of large sequencing centres including those at Whitehead Institute, the Sanger Centre, Washington University, and Baylor College of Medicine.

The Human Genome Project is considered a Mega Project because the human genome has approximately 3.3 billion base-pairs; if the cost of sequencing is US $3 per base-pair, then the approximate cost will be US $10 billion. If the sequence obtained was to be stored in book form, and if each page contained 1000 base-pairs recorded and each book contained 1000 pages, then 3300 such books would be needed in order to store the complete genome. However, if expressed in units of computer data storage, 3.3 billion base-pairs recorded at 2 bits per pair would equal 786 megabytes of raw data. This is comparable to a fully data loaded CD.

### *Public Versus Private Approaches*

In 1998, a similar, privately funded quest was launched by the American researcher Craig Venter, and his firm Celera Genomics. Venter was a scientist at the NIH during the early 1990s when the project was initiated. The $300,000,000 Celera effort was intended to proceed at a faster pace and at a fraction of the cost of the roughly $3 billion publicly funded project.

Celera used a technique called whole genome shotgun sequencing, employing pairwise end sequencing, which had been used to sequence bacterial genomes of up to six million base pairs in length, but not for anything nearly as large as the three billion base pair human genome.

Celera initially announced that it would seek patent protection on "only 200–300" genes, but later amended this to seeking "intellectual property protection" on "fully-characterized important structures" amounting to 100–300 targets. The firm eventually filed preliminary ("place-holder") patent applications on 6,500 whole or partial genes. Celera also promised to publish their findings in accordance with the terms of the 1996 "Bermuda Statement," by releasing new data annually (the HGP released its new data daily), although, unlike the publicly funded project, they would not permit free redistribution or scientific use of the data. The publicly funded competitor UC Santa Cruz was compelled to publish the first draft of the human genome before Celera for this reason. On July 7, 2000, the UCSC Genome Bioinformatics Group released a first working draft on the web. The scientific community downloaded one-half trillion bytes of information from the UCSC genome server in the first 24 hours of free and unrestricted access to the first ever assembled blueprint of our human species.

In March 2000, President Clinton announced that the genome sequence could not be patented, and should be made freely available to all researchers. The statement sent Celera's stock plummeting and dragged down the biotechnology-heavy Nasdaq. The biotechnology sector lost about $50 billion in market capitalization in two days.

Although the working draft was announced in June 2000, it was not until February 2001 that Celera and the HGP scientists published details of their drafts. Special issues of *Nature* (which published the publicly funded project's scientific paper) and *Science* (which published Celera's paper) described the methods used to produce the draft sequence and offered analysis of the sequence. These drafts covered about 83% of the genome (90% of the euchromatic regions with 150,000 gaps and the order and orientation of many segments not yet established). In February 2001, at the time of the joint publications, press releases announced that the project had been completed by both groups. Improved drafts were announced in 2003 and 2005, filling in to ≈ 92% of the sequence currently.

The competition proved to be very good for the project, spurring the public groups to modify their strategy in order to accelerate progress. The rivals at UC Santa Cruz initially agreed to pool their data, but the agreement fell apart when Celera refused to deposit its data in the unrestricted public database GenBank. Celera had incorporated the public data into their genome, but forbade the public effort to use Celera data. HGP is the most well known of many international genome projects aimed at sequencing the DNA of a specific organism. While the human DNA sequence offers the most tangible benefits, important developments in biology and medicine are predicted as a result of the sequencing of model organisms, including mice, fruit flies, zebrafish, yeast, nematodes, plants, and many microbial organisms and parasites.

In 2004, researchers from the International Human Genome Sequencing Consortium (IHGSC) of the HGP announced a new estimate of 20,000 to 25,000 genes in the human genome. Previously 30,000 to 40,000 had been predicted, while estimates at the start of the project reached up to as high as 2,000,000. The number continues to fluctuate and it is now expected that it will take many years to agree on a precise value for the number of genes in the human genome.

## History

In 1976, the genome of the RNA virus Bacteriophage MS2 was the first complete genome to be determined, by Walter Fiers and his team at the University of Ghent (Ghent, Belgium). The idea for the shotgun technique came from the use of an algorithm that combined sequence information from many small fragments of DNA to reconstruct a genome. This technique was pioneered by Frederick Sanger to sequence the genome of the Phage Φ-X174, a virus (bacteriophage) that primarily infects bacteria that was the first fully sequenced genome (DNA-sequence) in 1977. The technique was called shotgun sequencing because the genome was broken into millions of pieces as if it had been blasted with a shotgun. In order to scale up the method, both the sequencing and genome assembly had to be automated, as they were in the 1980s.

Those techniques were shown applicable to sequencing of the first free-living bacterial genome (1.8 million base pairs) of *Haemophilus influenzae* in 1995 and the first animal genome (~100 Mbp) It involved the use of automated sequencers, longer individual sequences using approximately 500 base pairs at that time. Paired sequences separated by a fixed distance of around 2000 base pairs which were critical elements enabling the development of the first genome assembly programs for reconstruction of large regions of genomes (aka 'contigs').

Three years later, in 1998, the announcement by the newly-formed Celera Genomics that it would scale up the pairwise end sequencing method to the human genome was greeted with skepticism in some circles. The shotgun technique breaks the DNA into fragments of various sizes, ranging from 2,000 to 300,000 base pairs in length, forming what is called a DNA "library". Using an automated DNA sequencer the DNA is read in 800bp lengths from both ends of each fragment. Using a complex genome assembly algorithm and a supercomputer, the pieces are combined and the genome can be reconstructed from the millions of short, 800 base pair fragments.

The success of both the public and privately funded effort hinged upon a new, more highly automated capillary DNA sequencing machine, called the Applied Biosystems 3700, that ran the DNA sequences through an extremely fine capillary tube rather than a flat gel. Even more critical was the development of a new, larger-scale genome assembly program, which could handle the 30–50 million sequences that would be required to sequence the entire human genome with this method.

At the time, such a program did not exist. One of the first major projects at Celera Genomics was the development of this assembler, which was written in parallel with the construction of a large, highly automated genome sequencing factory. Development of the assembler was led by Brian Ramos. The first version of this assembler was demonstrated in 2000, when the Celera team joined forces with

Professor Gerald Rubin to sequence the fruit fly Drosophila melanogaster using the whole-genome shotgun method. At 130 million base pairs, it was at least 10 times larger than any genome previously shotgun assembled. One year later, the Celera team published their assembly of the three billion base pair human genome.

The Human Genome Project was a 13 year old mega project, that was launched in the year 1990 and completed in 2003. This project is closely associated to the branch of biology called bioinformatics. The human genome project international consortium announced the publication of a draft sequence and analysis of the human genome—the genetic blueprint for the human being. An American company—Celera, led by Craig Venter and the other huge international collaboration of distinguished scientists led by Francis Collins, director, National Human Genome Research Institute, U.S., both published their findings.

This Mega Project is co-ordinated by the U.S. Department of Energy and the National Institute of Health. During the early years of the project, the Wellcome Trust (U.K.) became a major partner, other countries like Japan, Germany, China and France contributed significantly. Already the atlas has revealed some starting facts. The two factors that made this project a success are:

1. Genetic Engineering Techniques, with which it is possible to isolate and clone any segment of DNA.
2. Availability of simple and fast technologies, to determining the DNA sequences.

Being the most complex organisms, human beings were expected to have more than 100,000 genes or combination of DNA that provides commands for every characteristics of the body. Instead their studies show that humans have only 30,000 genes – around the same as mice, three times as many as flies, and only five times more than bacteria. Scientist told that not only are the numbers similar, the genes themselves, baring a few, are alike in mice and men.

In a companion volume to the Book of Life, scientists have created a catalogue of 1.4 million single-letter differences, or single-nucleotide polymorphisms (SNPs) – and specified their exact locations in the human genome. This SNP map, the world's largest publicly available catalogue of SNP's, promises to revolutionize both mapping diseases and tracing human history.

The sequence information from the consortium has been immediately and freely released to the world, with no restrictions on its use or redistribution. The information is scanned daily by scientists in academia and industry, as well as commercial database companies, providing key information services to bio-technologists. Already, many genes have been identified from the genome sequence, including more than 30 that play a direct role in human diseases.

By dating the three millions repeat elements and examining the pattern of interspersed repeats on the Y-chromosome, scientists estimated the relative mutation rates in the X and the Y chromosomes and in the male and the female germ lines. They found that the ratio of mutations in male Vs female is 2:1. Scientists point to several possible reasons for the higher mutation rate in the male germ line, including the fact that there are a greater number of cell divisions involved in the formation of sperm than in the formation of eggs.

### Methods

The IHGSC used pair-end sequencing plus whole-genome shotgun mapping of large ($\approx$ 100 Kbp) plasmid clones and shotgun sequencing of smaller plasmid sub-clones plus a variety of other mapping data to orient and check the assembly of each human chromosome. The Celera group emphasized the importance of the "whole-genome shotgun" sequencing method, relying on sequence information to orient and locate their fragments within the chromosome. However they used the publicly available data from HGP to assist in the assembly and orientation process, raising concerns that the Celera sequence was not independently derived.

### Genome Donors

In the IHGSC international public-sector Human Genome Project (HGP), researchers collected blood (female) or sperm (male) samples from a large number of donors. Only a few of many collected samples were processed as DNA resources. Thus the donor identities were protected so neither donors nor scientists could know whose DNA was sequenced. DNA clones from many different libraries were used in the overall project, with most of those libraries being created by Dr. Pieter J. de Jong. It has been informally reported, and is well known in the genomics community, that much of the DNA for the public HGP came from a single anonymous male donor from Buffalo, New York (code name RP11).

HGP scientists used white blood cells from the blood of two male and two female donors (randomly selected from 20 of each) — each donor yielding a separate DNA library. One of these libraries (RP11) was used considerably more than others, due to quality considerations.

One minor technical issue is that male samples contain just over half as much DNA from the sex chromosomes (one X chromosome and one Y chromosome) compared to female samples (which contain two X chromosomes). The other 22 chromosomes (the autosomes) are the same for both genders.

Although the main sequencing phase of the HGP has been completed, studies of DNA variation continue in the International HapMap Project, whose goal is to identify patterns of single-nucleotide polymorphism (SNP) groups (called haplotypes, or "haps"). The DNA samples for the HapMap came from a total of 270 individuals: Yoruba people in Ibadan, Nigeria; Japanese people in Tokyo; Han Chinese in Beijing; and the French Centre d'Etude du Polymorphisms Humain (CEf) resource, which consisted of residents of the United States having ancestry from Western and Northern Europe. In the Celera Genomics private-sector project, DNA from five different individuals were used for sequencing. The lead scientist of Celera Genomics at that time, Craig Venter, later acknowledged (in a public letter to the journal Science) that his DNA was one of 21 samples in the pool, five of which were selected for use.

On September 4, 2007, a team led by Craig Venter published his complete DNA sequence, unveiling the six-billion-nucleotide genome of a single individual for the first time.

### Benefits

The work on interpretation of genome data is still in its initial stages. It is anticipated that detailed knowledge of the human genome will provide new avenues for advances in medicine and biotechnology. Clear practical results of the project emerged even before the work was finished. For example, a number of companies, such as Myriad Genetics started offering easy ways to administer genetic tests that can show predisposition to a variety of illnesses, including breast cancer, disorders of hemostasis, cystic fibrosis, liver diseases and many others. Also, the etiologies for cancers, Alzheimer's disease and other areas of clinical interest are considered likely to benefit from genome information and possibly may lead in the long term to significant advances in their management.

There are also many tangible benefits for biological scientists. For example, a researcher investigating a certain form of cancer may have narrowed down his/her search to a particular gene. By visiting the human genome database on the World Wide Web, this researcher can examine what other scientists have written about this gene, including (potentially) the three-dimensional structure of its product, its function(s), its evolutionary relationships to other human genes, or to genes in mice or yeast or fruit flies, possible detrimental mutations, interactions with other genes, body tissues in which this gene is activated, diseases associated with this gene or other datatypes.

Further, deeper understanding of the disease processes at the level of molecular biology may determine new therapeutic procedures. Given the established importance of DNA in molecular biology and its central role in determining the fundamental operation of cellular processes, it is likely that expanded knowledge in this area will facilitate medical advances in numerous areas of clinical interest that may not have been possible without them.

The analysis of similarities between DNA sequences from different organisms is also opening new avenues in the study of evolution. In many cases, evolutionary questions can now be framed in terms of molecular biology; indeed, many major evolutionary milestones (the emergence of the ribosome and organelles, the development of embryos with body plans, the vertebrate immune system) can be related to the molecular level.

Many questions about the similarities and differences between humans and our closest relatives (the primates, and indeed the other mammals) are expected to be illuminated by the data from this project.

The Human Genome Diversity Project (HGDP), spinoff research aimed at mapping the DNA that varies between human ethnic groups, which was rumored to have been halted, actually did continue and to date has yielded new conclusions. In the future, HGDP could possibly expose new data in disease surveillance, human development and anthropology. HGDP could unlock secrets behind and create new strategies for managing the vulnerability of ethnic groups to certain diseases. It could also show how human populations have adapted to these vulnerabilities.

Advantages of Human Genome Project:

1. Knowledge of the effects of variation of DNA among individuals can revolutionize the ways to diagnose, treat and even prevent a number of diseases that affects the human beings.
2. It provides clues to the understanding of human biology.

### *Ethical, Legal and Social Issues*

The project's goals included not only identifying all of the approximately 24,000 genes in the human genome, but also to address the ethical, legal, and social issues (ELSI) that might arise from the availability of genetic information. Five percent of the annual budget was allocated to address the ELSI arising from the project.

Debra Harry, Executive Director of the U.S group Indigenous Peoples Council on Biocolonialism (IPCB), says that despite a decade of ELSI funding, the burden of genetics education has fallen on the tribes themselves to understand the motives of Human genome project and its potential impacts on their lives. Meanwhile, the government has been busily funding projects studying indigenous groups without any meaningful consultation with the groups.

The main criticism of ELSI is the failure to address the conditions raised by population-based research, especially with regard to unique processes for group decision-making and cultural worldviews. Genetic variation research such as HGP is group population research, but most ethical guidelines, according to Harry, focus on individual rights instead of group rights. She says the research represents a clash of culture: indigenous people's life revolves around collectivity and group decision making whereas the Western culture promotes individuality. Harry suggests that one of the challenges of ethical research is to include respect for collective review and decision making, while also upholding the Western model of individual rights.

## Cloning

Cloning in biology is the process of producing similar populations of genetically identical individuals that occurs in nature when organisms such as bacteria, insects or plants reproduce asexually. Cloning in biotechnology refers to processes used to create copies of DNA fragments (molecular cloning), cells (cell cloning), or organisms. The term also refers to the production of multiple copies of a product such as digital media or software.

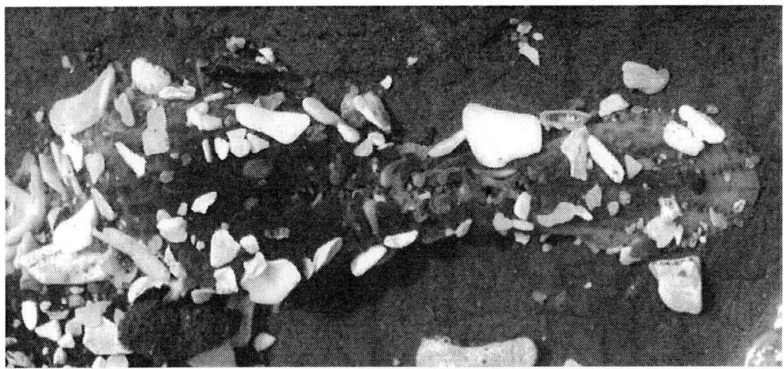

***Figure:*** *The sea anemone (Anthopleura elegantissima) undergoing cloning*

The Greek word for "trunk, branch", referring to the process whereby a new plant can be created from a twig. In horticulture, the spelling *clon* was used until the twentieth century; the final *e* came into use to indicate the vowel is a "long o" instead of a "short o". Since the term entered the popular lexicon in a more general context, the spelling *clone* has been used exclusively.

### Molecular Cloning

Molecular cloning refers to the procedure of isolating a defined DNA sequence and obtaining multiple copies of it *in vitro*. Cloning is frequently employed to amplify DNA fragments containing genes, but it can be used to amplify any DNA sequence such as promoters, non-coding sequences, chemically synthesised oligonucleotides and randomly fragmented DNA.

Cloning is used in a wide array of biological experiments and technological applications such as large scale protein production.

In essence, in order to amplify any DNA sequence *in vivo* and *in vitro*, the sequence in question must be linked to primary sequence elements capable of directing the replication and propagation of themselves and the linked sequence in the desired target host. The required sequence elements differ according to host, but invariably include an origin of replication, and a selectable marker.

In practice, however, a number of other features are desired and a variety of specialized cloning vectors exist that allow protein expression, tagging, single stranded RNA and DNA production and a host of other manipulations that are useful in downstream applications.

***Recombinase-based Cloning:*** A novel procedure of cloning or subcloning of any DNA fragment by inserting the special DNA fragment of interest into a special area of target DNA through interchange of the relevant DNA fragments.

This is a one-step reaction: simple, efficient, facilitating high throughput or automatic cloning and/or subcloning.

One of the currently popular recombinase-based systems is marketed under the name Gateway Technology.

***Restriction/Ligation Cloning:*** In the classical restriction and ligation cloning protocols, cloning of any DNA fragment essentially involves four steps: DNA fragmentation with restriction endonucleases, ligation of DNA fragments to a vector, transfection, and screening/ selection. Although these steps are invariable among cloning procedures a number of alternative routes can be selected at various points depending on the particular application; these are summarized as a 'cloning strategy'.

***Isolation of Insert:*** Initially, the DNA fragment to be cloned needs to be isolated. Preparation of DNA fragments for cloning can be accomplished in a number of alternative ways. Insert preparation is frequently achieved by means of polymerase chain reaction, but it may also be accomplished by restriction enzyme digestion, DNA sonication and fractionation by agarose gel electrophoresis. Chemically synthesized oligonucleotides can also be used if the target sequence size does not exceed the limit of chemical synthesis. Isolation of insert can be done by using shotgun cloning, c-DNA clones, gene machines (artificial chemical synthesis).

***Transformation:*** Following ligation, the ligation product (plasmid) is transformed into bacteria for propagation. The bacteria is then plated on selective agar to select for bacteria that have the plasmid of interest. Individual colonies are picked and tested for the wanted insert. Maxiprep can be done to obtain large quantity of the plasmid containing the inserted gene.

***Transfection:*** Following ligation, a portion of the ligation reaction, including vector with insert in the desired orientation is transfected into cells. A number of alternative techniques are available, such as chemical sensitization of cells, electroporation and biolistics.

Chemical sensitization of cells is frequently employed since this does not require specialized equipment and provides relatively high transformation efficiencies. Electroporation is used when extremely high transformation efficiencies are required, as in very inefficient cloning strategies. Biolistics are mainly utilized in plant cell transformations, where the cell wall is a major obstacle in DNA uptake by cells. The bacterial transformation is generally observed by blue white screening.

## Selection

Finally, the transfected cells are cultured. As the aforementioned procedures are of particularly low efficiency, there is a need to identify the cells that contain the desired insert at the appropriate orientation and isolate these from those not successfully transformed. Modern cloning vectors include selectable markers (most frequently antibiotic resistance markers) that allow only cells in which the vector, but not necessarily the insert, has been transfected to grow. Additionally, the cloning vectors may contain colour selection markers which provide blue/white screening (via $\alpha$-factor complementation) on X-gal medium.

Nevertheless, these selection steps do not absolutely guarantee that the DNA insert is present in the cells. Further investigation of the resulting colonies is required to confirm that cloning was successful. This may be accomplished by means of PCR, restriction fragment analysis and/or DNA sequencing.

### *Genetic Engineering*

Genetic engineering is a method of changing the inherited characteristics of an organism in a predetermined way by altering its genetic material. This is often done to enable micro-organisms, such as bacteria or viruses, to synthesize increased yields of compounds, to form entirely new compounds, or to adapt to different environments. Other uses of this technology, which is also called recombinant DNA technology, include gene therapy, which is the supply of a functional gene to a person with a genetic disorder or with other diseases such as acquired immune deficiency syndrome (AIDS) or cancer, and the cloning of whole organisms.

Genetic engineering involves the manipulation of deoxyribonucleic acid, or DNA. Important tools in this process are restriction endonucleases (so-called restriction enzymes) that are produced by various species of bacteria. Restriction enzymes can recognize a particular sequence of the chain of chemical units, called nucleotide bases, which make up the DNA molecule and cut the DNA at that location. Fragments of DNA generated in this way can be joined using other enzymes called ligases. Restriction enzymes and ligases therefore allow the specific cutting and reassembling of portions of DNA. Also important in the manipulation of DNA are so-called vectors, which are pieces of DNA that can self-replicate (produce copies of themselves) independently of the DNA in the host cell in which they are grown. Examples of vectors include plasmids, viruses, and artificial chromosomes. Vectors permit the generation of multiple copies of a particular piece of DNA, making this a useful method for generating sufficient quantities of material with which to work. The process of engineering a DNA fragment into a vector is called "molecular cloning", because multiple copies of an identical molecule of DNA are produced. Another way of producing many identical copies of a particular (often short, for example, 100-3,000 base pairs) DNA fragment is the polymerase chain reaction. This method is rapid and avoids the need for cloning DNA into a vector.

### *Gene Therapy*

Gene therapy involves supplying a functional gene to cells lacking that function, with the aim of correcting a genetic disorder or acquired disease. Gene therapy can be broadly divided into two categories. The first is alteration of germ cells, that is, sperm or eggs, which results in a permanent genetic change for the whole organism and subsequent generations. This "germ line gene therapy" is considered by many to be unethical in human beings. The second type of gene therapy, "somatic cell gene therapy", is analogous to an organ transplant. In this case, one or more specific tissues are targeted by direct treatment or by removal of the tissue, addition of the therapeutic gene or genes in the laboratory, and return of the treated cells to the patient. Clinical trials of somatic cell gene therapy began in the late 1990s, mostly for the treatment of cancers and blood, liver, and lung disorders.

The history of human gene therapy is, however, not a particularly happy one. The effect of introducing a gene into cells rarely promotes more than small transient relief from the symptoms of the disease being treated. Worse still, there have been highly publicized cases where gene therapy trial patients have suffered as a consequence of the treatment itself. For example, in 1999 an 18-year-old gene therapy trial volunteer from Philadelphia died following a gene therapy trial. In addition, one of the few success stories of human gene therapy—the treatment of severe combined immune deficiency, X-SCID—has turned out to have unforeseen consequences. Bone marrow cells were taken from patients suffering from this disease and treated with a virus to introduce a functional copy of the defective gene. When the modified bone marrow cells were returned to patients, their immune systems were functional once more. However, some patients treated this way subsequently developed leukaemia, which most likely arises as a result of random insertion of a section of DNA into the human genome with the consequent disruption of nearby gene function.

## Cloning Cells and Animals

In genetic engineering, the term "cloning" is now more commonly applied to the production of identical animals rather than molecular cloning of DNA fragments. Whole cell or animal cloning occurs through the transfer of the nucleus of an adult cell into an enucleated egg. This can result in the reprogramming of the adult cell DNA to produce a cloned animal. In 1997, a sheep named Dolly was born at the Roslin Institute in Edinburgh. She was created from the nucleus of a cultured mammary gland cell from a Finn Dorset sheep that was fused to an egg cell from a Scottish Blackface ewe that had had its own nucleus removed. The fused cell was implanted into a different Scottish Blackface ewe, and following a normal pregnancy, Dolly, a Finn Dorset sheep, was born. Nuclear transfer has subsequently been applied to produce a range of cloned animals including cows, goats, pigs, mice, and cats.

## Unicellular Organisms

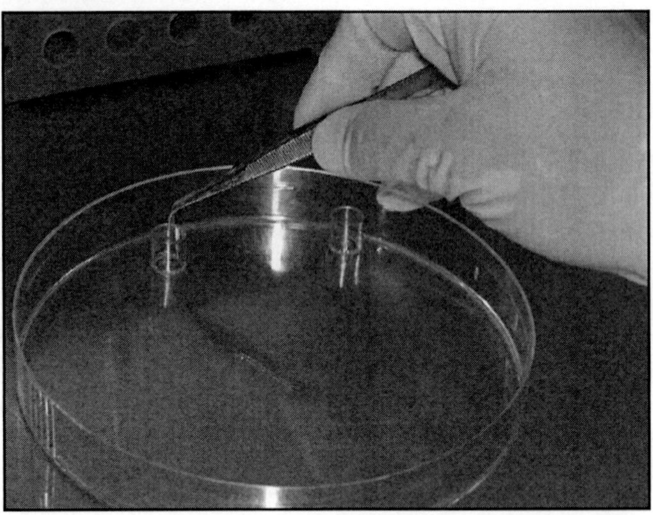

***Figure:*** *Cloning cell-line colonies using cloning rings*

Cloning a cell means to derive a population of cells from a single cell. In the case of unicellular organisms such as bacteria and yeast, this process is remarkably simple and essentially only requires the inoculation of the appropriate medium. However, in the case of cell cultures from multi-cellular organisms, cell cloning is an arduous task as these cells will not readily grow in standard media.

A useful tissue culture technique used to clone distinct lineages of cell lines involves the use of cloning rings (cylinders). According to this technique, a single-cell suspension of cells that have been exposed to a mutagenic agent or drug used to drive selection is plated at high dilution to create isolated colonies; each arising from a single and potentially clonal distinct cell. At an early growth stage when colonies consist of only a few of cells, sterile polystyrene rings (cloning rings), which have been dipped in grease are placed over an individual colony and a small amount of trypsin is added. Cloned cells are collected from inside the ring and transferred to a new vessel for further growth.

## Somatic Cell Nuclear Transfer

In genetics and developmental biology, somatic cell nuclear transfer (SCNT) is a laboratory technique for creating a clonal embryo, using an ovum with a donor nucleus. It can be used in embryonic stem cell research, or, potentially, in regenerative medicine where it is sometimes referred to as "therapeutic cloning." It can also be used as the first step in the process of reproductive cloning.

### *The Process*

In SCNT the nucleus, which contains the organism's DNA, of a somatic cell (a body cell other than a sperm or egg cell) is removed and the rest of the cell discarded. At the same time, the nucleus of an egg cell is removed. The nucleus of the somatic cell is then inserted into the denucleated egg cell. After being inserted into the egg, the somatic cell nucleus is reprogrammed by the host cell. The egg, now containing the nucleus of a somatic cell, is stimulated with a shock and will begin to divide. After many mitotic divisions in culture, this single cell forms a blastocyst (an early stage embryo with about 100 cells) with almost identical DNA to the original organism. The technique of transferring a nucleus from a somatic cell into an egg that produced Dolly was an extension of experiments that had been ongoing for over 40 years. In the simplest terms, the technique used to produce Dolly the sheep-somatic cell nuclear transplantation cloning-involves removing the nucleus of an egg and replacing it with the diploid nucleus of a somatic cell.

### *SCNT in Stem Cell Research*

Some researchers use SCNT in stem cell research. The aim of carrying out this procedure is to obtain stem cells that are genetically matched to the donor organism. Presently, no human stem cell lines have been derived from SCNT research. Embryonic stem cells are new, unspecialized cells that are able to be produced into a specialized cell that can replace another cell that has been lost in the body.

A potential use of genetically-customized stem cells would be to create cell lines that have genes linked to the particular disease. For example, if a person with Parkinson's disease donated his or her somatic cells, then the stem cells resulting from SCNT would have genes that contribute to Parkinson's disease. In this scenario, the disease specific stem cell lines would be studied in order to better understand the disease. In another scenario, genetically-customized stem cell lines would be generated for cell-based therapies to transplant to the patient. The resulting cells would be genetically identical to the somatic cell donor, thus avoiding any complications from immune system rejection.

Only a handful of the labs in the world are currently using SCNT techniques in human stem cell research. In the United States, scientists at the Harvard Stem Cell Institute, the University of California San Francisco, Stemagen (La Jolla, CA) and possibly Advanced Cell Technology are currently researching a technique to use somatic cell nuclear transfer to produce embryonic stem cells. In the United Kingdom, the Human Fertilisation and Embryology Authority has granted permission to research groups at the Roslin Institute and the

Newcastle Centre for Life. SCNT may also be occurring in China. In 2005, a South Korean research team led by Professor Hwang Woo-suk, published claims to have derived stem cell lines via SCNT, but supported those claims with fabricated data. Recent evidence has proved that he in fact created a stem cell line from a parthenote.

The impetus for SCNT-based stem cell research has been decreased by the development and improvement of alternative methods of generating stem cells. Methods to reprogram normal body cells into pluripotent stem cells were developed in humans in 2007. The following year, this method achieved a key goal of SCNT-based stem cell research: the derivation of pluripotent stem cell lines that have all genes linked to various diseases. Some scientists working on SCNT-based stem cell research have recently moved to the new methods of induced pluripotent stem cells.

### SCNT in Reproductive Cloning

This technique is currently the basis for cloning animals (such as the famous Dolly the sheep), and in theory could be used to clone humans. However, most researchers believe that in the foreseeable future it will not be possible to use this technique to produce a human clone that will develop to term. However, it is still a possibility and can become more probable in the future as it will probably need a few more adjustments to work for humans.

### Limitations

Stresses placed on both the egg cell and the introduced nucleus are enormous, leading to a high loss in resulting cells. For example, Dolly the sheep was born after 277 eggs were used for SCNT, which created 29 viable embryos. Only three of these embryos survived until birth, and only one survived to adulthood. As the procedure currently cannot be automated, but has to be performed manually under a microscope, SCNT is very resource intensive. The biochemistry involved in reprogramming the differentiated somatic cell nucleus and activating the recipient egg is also far from understood. In SCNT, not all of the donor cell's genetic information is transferred, as the donor cell's mitochondria that contain their own mitochondrial DNA are left behind. The resulting hybrid cells retain those mitochondrial structures which originally belonged to the egg. As a consequence, clones such as Dolly that are born from SCNT are not perfect copies of the donor of the nucleus.

### Controversy

Proposals to use nucleus transfer techniques in human stem cell research raise a set of concerns beyond the moral status of any created embryo. These have led to some individuals and organizations who are *not* opposed to human embryonic stem cell research to be concerned about, or opposed to, SCNT research.

One concern is that blastula creation in SCNT-based human stem cell research will lead to the reproductive cloning of humans. Both processes use the same first step: the creation of a nuclear transferred embryo, most likely via SCNT. Those who hold this concern often advocate for strong regulation of SCNT to preclude implantation of any derived products for the intention of human reproduction, or its prohibition.

A second important concern is the appropriate source of the eggs that are needed. SCNT requires human eggs, which can only be obtained from women. The most common source of these eggs today are eggs that are produced and in excess of the clinical need during IVF treatment. This is a minimally invasive procedure, but it does carry some health risks, such as ovarian hyperstimulation syndrome, and in very rare instances even death.

One vision for successful stem cell therapies is to create custom stem cell lines for patients. Each custom stem cell line would consist of a collection of identical stem cells each carrying the patient's own DNA, thus reducing or eliminating any problems with rejection when the stem cells were transplanted for treatment. For example, to treat a man with Parkinson's disease, a cell nucleus from one of his cells would be transplanted by SCNT into an egg cell from an egg donor, creating a unique lineage of stem cells almost identical to the patient's own cells. (There would be differences. For example, the mitochondrial DNA would be the same as that of the egg donor. In comparison, his own cells would carry the mitochondrial DNA of his mother.)

Potentially millions of patients could benefit from stem cell therapy, and each patient would require a large number of donated eggs in order to successfully create a single custom therapeutic stem cell line. Such large numbers of donated eggs would exceed the number of eggs currently left over and available from couples trying to have children through assisted reproductive technology.

Therefore, healthy young women would need to be induced to sell eggs to be used in the creation of custom stem cell lines that could then be purchased by the medical industry and sold to patients. It is so far unclear where all these eggs would come from. The sale of human eggs is normally referred to as a "donation," but women who donate their eggs are often paid thousands of dollars.

Stem cell experts consider it unlikely that such large numbers of human egg donations would occur in developed country because of the unknown long-term public health effects of treating large numbers of healthy young women with heavy doses of hormones in order to induce hyperovulation (ovulating several eggs at once). Although such treatments have been performed for several decades now, the long-term effects have not been studied or declared safe to use on a large scale on otherwise healthy women. Longer-term treatments with much lower doses of hormones are known to increase the rate of cancer decades later. Whether hormone treatments to induce hyperovulation could have similar effects is unknown. There are also ethical questions surrounding paying for eggs. In general, marketing body parts is considered unethical and is banned in most countries. Human eggs have been a notable exception to this rule for some time.

To address the problem of creating a human egg market, some stem cell researchers are investigating the possibility of creating artificial eggs. If successful, human egg donations would not be needed to create custom stem cell lines. However, this technology may be a long way off.

### *Policies Regarding Human SCNT*

SCNT involving human cells is currently legal for research purposes in the United Kingdom, having been incorporated into the Human Fertilisation and Embryology Act 1990 in 2001. Permission must be obtained from the Human Fertilisation and Embryology Authority in order to perform or attempt SCNT.

In the United States, the practice remains legal, as it has not been addressed by federal law. However, in 2002, a moratorium on United States federal funding for SCNT prohibits funding the practice for the purposes of research. Thus, though legal, SCNT cannot be federally funded. In 2003, the United Nations adopted a proposal submitted by Costa Rica, calling on member states to "prohibit all forms of human cloning in as much as they are incompatible with human dignity and the protection of human life." This phrase may include SCNT, depending on interpretation.

The Council of Europe's *Convention on Human Rights and Biomedicine* and its *Additional Protocol to the Convention for the Protection of Human Rights and Dignity of the Human Being with regard to the Application of Biology and Medicine, on the Prohibition of Cloning Human Being* appear to ban SCNT of human beings. Of the Council's 45 member states, the *Convention* has been signed by 31 and

The Grocery Manufacturers of America estimate that 75% of all processed foods in the U.S. contain a GM ingredient.

In particular, Bt corn, which produces the pesticide within the plant itself, is widely grown, as are soybeans genetically designed to tolerate glyphosate herbicides. These constitute "input-traits" are aimed to financially benefit the producers, have indirect environmental benefits and marginal cost benefits to consumers.

In the US, by 2009/2010, 93% of the planted area of soybeans, 93% of cotton, 86% of corn and 95% of the sugar beet were genetically modified varieties. Genetically modified soybeans carried herbicide-tolerant traits only, but maize and cotton carried both herbicide tolerance and insect protection traits (the latter largely the *Bacillus thuringiensis* Bt insecticidal protein). In the period 2002 to 2006, there were significant increases in the area planted to Bt protected cotton and maize, and herbicide tolerant maize also increased in sown area.

## Genetically Modified Food Controversies

The genetically modified foods controversy is a dispute over the relative advantages and disadvantages of genetically modified (GM) food crops and other uses of genetically-modified organisms in food production. The dispute involves biotechnology companies, governmental regulators, non-governmental organizations and scientists. The dispute is most intense in Japan and Europe, where public concern about GM food is higher than in other parts of the world such as the United States. In the United States GM crops are more widely grown and the introduction of these products has been less controversial.

The five key areas of political controversy related to genetically engineered food are food safety, the effect on natural ecosystems, gene flow into non GE crops, moral/religious concerns, and corporate control of the food supply. To date, not a single instance of harm to human health or the environment has been documented with GM crops. Several benefits have been widely accepted and are uncontested in the scientific literature. These include reductions in insecticide use on GE cotton, enhanced biological diversity in GE cotton fields (compared to non-GE fields, enhanced farmer income and communal benefits, increased yields for poor farmers and improved health of farmworkers. Although the use of herbicide tolerant crops remain controversial, because of the need to spray herbicides, it is clear that the use of these crops has promoted a shift to less toxic herbicides.

The concept of substantial equivalence has been established to address the fears of some consumers about the safety of GE crops. "Substantial equivalence embodies the concept that if a new food or food component is found to be substantially equivalent to an existing food or food component, it can be treated in the same manner with respect to safety (i.e., the food or food component can be concluded to be as safe as the conventional food or food component)". The rationale for this approach is that it would be impossible to test all the new crop varieties every year for food safety. Only a few food crops on the market have been shown to cause adverse health effects and all of these were the result of conventional genetic modification, not from genetic engineering To date no adverse health effects caused by products approved for sale have been documented, although two products failed initial safety testing and were discontinued, due to allergic reactions. Most feeding trials have observed no toxic effects and saw that GM foods were equivalent in nutrition to unmodified foods, although a few non-peer-reviewed reports speculate physiological changes to GM food. Although there is now broad scientific consensus that GE crops on the market are safe to eat, some scientists and advocacy groups such as Greenpeace and World Wildlife Fund call for additional and more rigorous testing before marketing genetically engineered food.

Another area of controversy is what effect pest and herbicide-resistant crops have on ecosystems. For example, whereas 14 years of studies indicate cotton engineered for resistance to insects has resulted in massive reduction in insecticides globally, the use of crops engineered for herbicide tolerance has changed herbicide patterns, with a clear shift to use of less toxic herbicides. Studies assessing diversity of non-target insects has revealed enhanced biological diversity in GE fields compared to non-GE fields. The risk and effects of have also been cited as concerns, with the possibility that genes might spread from modified crops to wild relatives. There is no evidence that the risk of pollen transfer is greater for any of the GE crops on the market as compared to non-GE crops.

## Health Risks and Benefits

### *Present Knowledge on GM Food Safety*

Worldwide, there is a range of perspectives within non-governmental organizations on the safety of GM foods. For example, the US pro-GM group AgBioWorld has argued that GM foods have been proven safe, while other pressure groups and consumer rights groups, such as the Organic Consumers Association, and Greenpeace claim the long-term health risks which GM could pose, or the environmental risks associated with GM, have not yet been adequately investigated. In Japan, Consumers Union of Japan are opposed to GMO foods. They also claim that truly independent research in these areas is systematically blocked by the GM corporations which own the GM seeds and reference materials.

A 2008 review published by the Royal Society of Medicine noted that GM foods have been eaten by millions of people worldwide for over 15 years, with no reports of ill effects. Similarly a 2004 report from the US National Academies of Sciences stated: "To date, no adverse health effects attributed to genetic engineering have been documented in the human population." A 2004 review of feeding trials in the *Italian Journal of Animal Science* found no differences among animals eating genetically modified plants. A 2005 review in *Archives of Animal Nutrition* concluded that first-generation genetically modified foods had been found to be similar in nutrition and safety to non-GM foods, but noted that second-generation foods with "significant changes in constituents" would be more difficult to test, and would require further animal studies. However, a 2009 review in *Nutrition Reviews* found that although most studies concluded that GM foods do not differ in nutrition or cause any detectable toxic effects in animals, some studies did report adverse changes at a cellular level caused by some GM foods, concluding that "More scientific effort and investigation is needed to ensure that consumption of GM foods is not likely to provoke any form of health problem". A review published in 2009 by Dona and Arvanitoyannis concluded that "results of most studies with GM foods indicate that they may cause some common toxic effects such as hepatic, pancreatic, renal, or reproductive effects and may alter the hematological, biochemical, and immunologic parameters".

However responses to this review in 2009 and 2010 note that the Dona and Arvanitoyannis concentrated on articles with an anti-GM bias that have been refuted by scientists in peer-reviewed articles elsewhere- for example the 35S promoter, stability of transgenes, antibiotic marker genes and the claims for toxic effects of GM foods. In 2007, a review by Domingo of the toxicity by searching in the Publimed data base using 12 search terms, cited 68 references, found that the "number of references" on the safety of GM/transgenic crops was "surprisingly limited" and questioned whether the safety of genetically modified food has been demonstrated; the review also remarked that its conclusions were in agreement with three earlier reviews by Zdunczyk (2001), Bakshi (2003), and Pryme and Lembcke (2003).

However, an article in 2007 by Vain found 692 research studies focusing on GM crop and food safety and identified a strong increase in the publication of such articles in recent years. Vain commented that the multidisciplinarian nature of GM research complicates the retrieval of GM studies and requires using many search terms (he used more than 300) and multiple data bases.

### Safety Assessments

The starting point for the safety assessment of genetically engineered food products is to assess if the food is "substantially equivalent" to its natural counterpart. To decide if a modified product is substantially equivalent, the product is tested by the manufacturer for unexpected changes in a limited set of components such as toxins, nutrients or allergens that are present in the unmodified food. If these tests show no significant difference between the modified and unmodified products, then no further food safety testing is required. The manufacturers data is then assessed by an independent regulatory body, such as the U.S. Food and Drug Administration.

However, if the product has no natural equivalent, or shows significant differences from the unmodified food, then further safety testing is carried out. A 2003 review in *Trends in Biotechnology* identified 7 main parts of a standard safety test:

1. Study of the introduced DNA and the new proteins or metabolites that it produces;
2. Analysis of the chemical composition of the relevant plant parts, measuring nutrients, anti-nutrients as well as any natural toxins or known allergens;
3. Assess the risk of gene transfer from the food to microorganisms in the human gut;
4. Study the possibility that any new components in the food might be allergens;
5. Estimate how much of a normal diet the food will make up;
6. Estimate any toxicological or nutritional problems revealed by this data;
7. Additional animal toxicity tests if there is the possibility that the food might pose a risk.

This process was examined further in a review published by Kuiper *et al.* 2002 in the journal *Toxicology*, which stated that substantial equivalence does not itself measure risks, but instead identifies differences between existing products and new foods, which might pose dangers to health. If differences do exist, identifying these differences is a starting point for a full safety assessment, rather than an end point. The authors concluded that "The concept of substantial equivalence is an adequate tool in order to identify safety issues related to genetically modified products that have a traditional counterpart". However, the review also noted difficulties in applying this standard in practice, including the fact that traditional foods contain many chemicals that have toxic or carcinogenic effects and that our existing diets therefore have not been proven to be safe. This lack of knowledge on unmodified food poses a problem, as GM foods may have differences in anti-nutrients and natural toxins that have never been identified in the original plant, raising the possibility that harmful changes could be missed.

The application of substantial equivalence has also been more strongly criticized. For example, in a speech in 1999, Andrew Chesson of the University of Aberdeen, stated that substantial equivalence testing "could be flawed in some cases" and that some current safety tests could allow harmful substances to enter the human food chain. In a commentary in *Nature* Millstone *et al.* argued that all GM foods should have extensive biological, toxicological and immunological tests and that the concept of substantial equivalence based solely on chemical analyses of the components of a food should be abandoned. They stated that this is necessary since it is currently impossible to predict the biological properties of a substance only from knowledge of its chemistry. This commentary was controversial and was criticized for misleading presentation of data and presenting an over-simplified version of safety assessments. For example, Kuiper *et al.* responded to this criticism by noting that equivalency testing does involve more than chemical tests and may include toxicity testing.

Medical writer Barbara Keeler and Marc Lappe argued in a 2001 article in the Los Angeles Times that the differences between Genetically Modified and conventional foods challenges the presumption of equivalence. Using Roundup ready soy that has been on the market since 1995 as an example, they noted the differences when compared to its unmodified counterpart.

Significantly lower levels of protein than unmodified soy. Significantly lower levels of phenylalanine, an essential amino acid and as a dietary supplement, the reason doctors advise the consumption of soy products. Levels of trypsin inhibitor were 27% higher and after toasting, Lectin was double that found in conventional soy, both are known allergens. GM soy also has 29% less choline, a B-complex vitamin. Round up ready soy had also stunted the growth of rats in Monsanto's study but had not effected cattle although it had increased the fat content of their milk. The authors do not maintain that modified soy is a hazard but that the FDA accepting such significant differences as being *substantially equivalent* illustrates the need for more rigorous testing, and preferably not by the biotech industries themselves.

The value of current independent studies is problematic as, due to restrictive End-user agreements, researchers are forbidden by law from publishing independent research in peer reviewed journals without the approval of the agritech companies. Cornell University's Elson Shields, the spokesperson for a group of scientists who oppose this practice, submitted a statement to the United States Environmental Protection Agency (EPA) protesting that "as a result of restrictive access, no truly independent research can be legally conducted on many critical questions regarding the technology".

Scientific American noted that several studies that were initially approved by seed companies were later blocked from publication when they returned "unflattering" results. While recognising that seed companies' intellectual property rights need to be protected, Scientific American calls the practice dangerous and has called for the restrictions on research in the End-user agreements to be lifted immediately and for the EPA to require, as a condition of approval, that independent researchers have unfettered access to GM products for testing.

The Welsh pressure group GM Free Cymru argues that governments should use independent studies rather than industry studies to assess crop safety. GM Free Cymru has also stated that independently funded researcher, Professor Bela Darvas of Debrecen University was refused Mon 863 Bt corn to use in his studies after previously publishing that a different variety of Monsanto corn was lethal to two Hungarian protected insect species and an insect classified as a rare.

### Allergenicity

Worldwide, reports of allergies to all kinds of foods, particularly nuts, fish and shellfish, seem to be increasing, but it is not known if this reflects a genuine change in the risk of allergy, or an increased awareness of food allergies by the public. Some environmental organizations, such as the European Green Party and Greenpeace, have suggested that GM food might trigger food allergies, although other environmentalists have implicated causes as diverse as the greenhouse effect increasing pollen levels, greater exposure to synthetic chemicals, cleaner lifestyles, or more mold in buildings. A 2005 review in the journal *Allergy* of the results from allergen testing of current GM foods stated that "no biotech proteins in foods have been documented to cause allergic reactions".

A well-known case of a GM plant that did not reach the market due to it producing an allergic reaction was a new form of soybean intended for animal feed. The allergen was transferred unintentionally from the Brazil nut into genetically engineered soybeans, in a bid to improve soybean nutritional quality for animal feed use. This new protein increased the levels in the GM soybean of the natural essential amino acid methionine, which is commonly added to poultry feed. Investigation of the GM soybeans revealed that they produced immune reactions in people with Brazil nut allergies, since the methionine rich protein chosen by Pioneer Hi-Bred happened to be a major source of Brazil nut allergy. Although this soybean strain was not developed as a human food, Pioneer Hi-Bred discontinued further development of the GM soybean, due to the difficulty in ensuring that none of these soybeans entered the human food chain.

In November 2005 a pest-resistant field pea developed by the Australian CSIRO for use as a pasture crop was shown to cause an allergic reaction in mice. Work on this variety was immediately halted. The protein added to the pea did not cause the reaction in humans or mice in isolation, but when it was expressed in the pea, it exhibited a subtly different structure which may have caused the allergic reaction. The immunologist who tested the pea noted that crops need to be evaluated case-by-case, and a Greenpeace activist commented that to his knowledge, no countries required feeding studies for approval of genetically modified foods.

Plant scientist Maarten J Chrispeels has made these comments about this example:

The recent Prescott et al. paper in JFAC contains a very interesting study on the immunogenicity of amylase [starch digestion enzyme] inhibitor in its native form (isolated from beans) and expressed as a transgene in peas. First of all, amylase inhibitor is a food protein, but also a "toxic" protein because it inhibits our digestive amylases. This is one of the reasons you have to cook your beans! (The other toxic bean protein is phytohemagglutinin and it is much more toxic).

This particular amylase inhibitor is found in the common bean (other species have other amylase inhibitors). Even though it is a food protein, it is unlikely ever to be used for genetic engineering of human foods because it inhibits our amylases. What the results show is that the protein, when synthesized in pea cotyledons has a different immunogenicity than when it is isolated from bean cotyledons (the native form). This is somewhat surprising but may be related to the presence of slightly different carbohydrate chains.

These cases of products that failed safety testing can either be viewed as evidence that genetic modification can produce unexpected and dangerous changes in foods, or alternatively that the current tests are effective at identifying any safety problems before foods come on the market.

Genetic modification can also be used to remove allergens from foods, which may, for example, allow the production of soy products that would pose a smaller risk of food allergies than standard soybeans. A hypo-allergenic strain of soybean was tested in 2003 and shown to lack the major allergen that is found in the beans. A similar approach has been tried in ryegrass, which produces pollen that is a major cause of hayfever: here a fertile GM grass was produced that lacked the main pollen allergen, demonstrating that the production of hypoallergenic grass is also possible.

### Environmental Risks and Benefits

The large scale growth of GM plants may have both positive and negative effects on the environment. These may be both direct effects, on organisms that feed on or interact with the crops, or wider effects on food chains produced by increases or decreases in the numbers of other organisms. As an example of benefits, insect-resistant Bt-expressing crops will reduce the number of pest insects feeding on these plants, but as there are fewer pests, farmers do not have to apply as much insecticide, which in turn tends to increase the number of non-pest insects in these fields. A 2006 study of the global impact of GM crops, published by the UK consultancy PG Economics and funded by Illinois-Missouri Biotechnology Alliance, concluded that globally, the technology reduced pesticide spraying by 286,000 tons in 2006, decreasing the environmental impact of herbicides and pesticides by 15%.

By reducing the amount of ploughing needed, GM technology led to reductions of greenhouse gases from soil equivalent to removing 6.56 million cars from the roads. However, a 2009 study published by the Organic Centre stated that the use of genetically engineered corn, soybean, and cotton increased the use of herbicides by 383 million pounds (191,500 tons), and pesticide use by 318.4 million pounds (159,200 tons). As an example of a concern about environmental risk, a lab at Cornell University published an article which caused worry in

The United Methodist Church opposed research and reproductive cloning in May 2000 and again in May 2004.

***Jewish:*** Judaism does not equate life with conception and, though some question the wisdom of cloning, Orthodox rabbis generally find no firm reason in Jewish law and ethics to object to cloning. Liberal Jewish thinkers have cautioned against cloning, among other genetic engineering efforts, though some eye the potential medical advantages.

***Buddhism:*** Ronald Y. Nakasone, a Buddhist priest and Professor of Buddhist Art and Culture at the Graduate Theological Union in Berkeley, California stated, "The Buddhist response to the possibility of cloning human beings is not if, but when... Would we accord a cloned person the benefits enjoyed by those who are born naturally? I would hope so."

***Raëlian:*** Raëlism is the only religious group of which any part (specifically, the religion's medical arm Clonaid) has claimed to have successfully cloned a human being. Clonaid claims that cloning will bring humanity closer to immortality.

Following the announcement, then-White House Press Secretary Scott McClellan spoke on behalf of president George W. Bush and said that human cloning was "deeply troubling" to most Americans. Kansas Republican Sam Brownback said that Congress should ban all human cloning, while some Democrats were worried that Clonaid announcement would lead to the banning of therapeutic cloning. FDA biotechnology chief Dr. Phil Noguchi warned that the human cloning, even if it worked, risked transferring sexually transmitted diseases to the newly born child. Clonaid claimed that it had a list of couples who were ready to have a cloned child.

University of Wisconsin–Madison biocthicist Alta Charo said that even in other ape-like mammals, the risk for miscarriage, birth defects, and life problems remains high. Robert Lanza of Advanced Cell Technologies said that Clonaid has no record of accomplishment for cloning anything, but he said that if Clonaid actually succeeded, there would be public unrest that may lead to the banning of therapeutic cloning, which has the capacity to cure millions of patients. The Vatican said that the claims expressed a mentality that was brutal and lacked ethical consideration. The White House was also critical of the claims.

### Governmental Actions

On December 28, 2006, the U.S. Food and Drug Administration (FDA) approved the consumption of meat and other products from cloned animals. Cloned-animal products were said to be virtually indistinguishable from the non-cloned animals. ANTONIO companies would not be required to provide labels informing the consumer that the meat comes from a cloned animal. In 2007, some meat and dairy producers did propose a system to track all cloned animals as they move through the food chain, suggesting that a national database system integrated into the National Animal Identification System could eventually allow CLETUS food labelling. However, no tracking system currently exists, and products from the offspring of cloned animals are increasingly sold for human consumption in the United States.

Critics have raised objections to the FDA's approval of cloned-animal products for human consumption, arguing that the FDA's research was inadequate, inappropriately limited, and of questionable scientific validity. Several consumer-advocate groups are working to encourage a tracking program that would allow consumers to become more aware of cloned-animal products within their food.

### Current Law

***United Nations:*** On December 14, 2001, the United Nations General Assembly began elaborating an international convention against the reproductive cloning of humans. A broad coalition of States, including Spain, Italy, Philippines, the United States, Costa Rica and the Holy See sought to extend the debate to ban all forms of human cloning, noting that, in their view, therapeutic human cloning violates human dignity. Costa Rica proposed the adoption of an international convention to ban all forms of Human Cloning. Unable to reach a consensus on a binding convention, in March 2005 a non-binding *United Nations Declaration on Human Cloning* calling for the ban of all forms of Human Cloning contrary to human dignity, was finally adopted.

***Australia:*** Australia had prohibited human cloning, though as of December 2006, a bill legalising therapeutic cloning and the creation of human embryos for stem cell research passed the House of Representatives. Within certain regulatory limits, and subject to the effect of state legislation, therapeutic cloning is now legal in some parts of Australia.

***European Union:*** The European Convention on Human Rights and Biomedicine prohibits human cloning in one of its additional protocols, but this protocol has been ratified only by Greece, Spain and Portugal. The Charter of Fundamental Rights of the European Union explicitly prohibits reproductive human cloning. The charter is legally binding for the institutions of the European Union under the Treaty of Lisbon.

***United States:*** In 1998, 2001, 2004 and 2007, the United States House of Representatives voted whether to ban all human cloning, both reproductive and therapeutic. Each time, divisions in the Senate over therapeutic cloning prevented either competing proposal (a ban on both forms or reproductive cloning only) from passing. On Mar 10, 2010 a bill (HR 4808) was introduced with a section banning federal funding for human cloning. Such a law, if passed, would not prevent research from occurring in private institutions (such as universities) that have both private and federal funding. There are currently no federal laws in the United States which ban cloning completely, and any such laws would raise difficult Constitutional questions similar to the issues raised by abortion. Thirteen American states (AR, CA, CT, IA, IN, MA, MD, MI, ND, NJ, RI, SD, VA) ban reproductive cloning and three states (AZ, MD, MO) prohibit use of public funds for such activities.

***United Kingdom:*** On January 14, 2001 the British government passed The Human Fertilisation and Embryology (Research Purposes) Regulations 2001 to amend the Human Fertilisation and Embryology Act 1990 by extending allowable reasons for embryo research to permit research around stem cells and cell nuclear replacement, thus allowing therapeutic cloning. However, on 15 November 2001, a pro-life group won a High Court legal challenge, which struck down the regulation and effectively left all forms of cloning unregulated in the UK. Their hope was that Parliament would fill this gap by passing prohibitive legislation. Parliament was quick to pass Human Reproductive Cloning Act 2001 which explicitly prohibited reproductive cloning. The remaining gap with regard to therapeutic cloning was closed when the appeals courts reversed the previous decision of the High Court.

The first licence was granted on August 11, 2004 to researchers at the University of Newcastle to allow them to investigate treatments for diabetes, Parkinson's disease and Alzheimer's disease. The Human Fertilisation and Embryology Act 2008, a major review of fertility legislation, repealed the 2001 Cloning Act by making amendments of similar effect to the 1990 Act. The 2008 Act also allows experiments on hybrid human-animal embryos.

### In Popular Culture

Cloning is a recurring theme in contemporary science fiction. Examples include the novels *Joshua Son of None* (about the cloning of an assassinated U.S. President strongly implied to be John F. Kennedy), *The Boys from Brazil* (cloning Adolf Hitler) and *A Parade of Mirrors and Reflections* by Anatoly Kudryavitsky (cloning Yuri Andropov). The

*Star Wars* films, TV series *The Clone Wars*, the animated series Clone High, as well as the 2000 Arnold Schwarzenegger film *The 6th Day* and 2005 *The Island*, directed by Michael Bay, also explore the theme of human cloning. 2005 *Æon Flux* depicts a future when the whole human species survives by means of cloning due to generalized infertility. An episode of Star Trek: Enterprise (Similitude) deals with the moral and ethical issues surrounding growing a human clone to harvest tissue for an injured crewman.

The film *Womb* deals with these issues with respect to death of a beloved person in a private relationship.

The 2001 Brazilian telenovela *O Clone* (as well as its 2010 remake *El Clon*) has human cloning as the main plot.

The famous video game franchise Metal Gear Solid, also revolves around the concept of cloning and genetic alteration. In Margaret Peterson Haddix's novel Double Identity, Bethany is an exact copy of her deceased older sister Elizabeth. The young adult science fiction novel The House of the Scorpion, by Nancy Farmer, also explores the idea of cloning. In *The Super Dimension Fortress Macross* (1982) anime series the Earth is attacked by an alien humanoid race of giants called *Zentradi* who are reproduced by cloning. This series was adapted years later into the first part of Robotech (1985), where the aliens remained the same but had a different origin. In the episode *The Doctor's Daughter* of BBC Television's long-running science fiction series *Doctor Who*, a tissue sample from The Doctor's arm is used to create a full-grown female soldier (whom The Doctor is both biological mother and father of) ready to fight. Diploid cells in The Doctor's tissue sample were split into Haploid cells, and then combined in a different arrangement and grown at a fast rate, a process which The Doctor calls "Progenation".

Human cloning also gained a foothold in popular culture, starting in the 1970s. Alvin Toffler's *Future Shock*, David Rorvik's *In his Image: The Cloning of a Man*, Woody Allen's film *Sleeper* and *The Boys from Brazil* all helped to make the public aware of the ethical issues surrounding human cloning.

In the Superman canon the comic "Krypton" delves into the ethics and effects of cloning. Kryptonians each have three genetic copies in case they need body parts. This starts a war for clone rights.

Aldous Huxley's novel "Brave New World"(1932) envisioned a world where large numbers of human clones would be cultivated industrially and conditioned before "birth" for specific castes.

ratified by 18. The *Additional Protocol* has been signed by 29 member nations and ratified by 14.

### Organism Cloning

Organism cloning (also called reproductive cloning) refers to the procedure of creating a new multicellular organism, genetically identical to another. In essence this form of cloning is an asexual method of reproduction, where fertilization or inter-gamete contact does not take place. Asexual reproduction is a naturally occurring phenomenon in many species, including most plants and some insects. Scientists have made some major achievements with cloning, including the asexual reproduction of sheep and cows. There is a lot of ethical debate over whether or not cloning should be used. However, cloning, or asexual propagation, has been common practice in the horticultural world for hundreds of years.

### Horticultural

The term *clone* is used in horticulture to mean all descendants of a single plant, produced by vegetative reproduction or apomixis. Many horticultural plant cultivars are clones, having been derived from a single individual, multiplied by some process other than sexual reproduction. As an example, some European cultivars of grapes represent clones that have been propagated for over two millennia. Other examples are potato and banana. Grafting can be regarded as cloning, since all the shoots and branches coming from the graft are genetically a clone of a single individual, but this particular kind of cloning has not come under ethical scrutiny and is generally treated as an entirely different kind of operation. Many trees, shrubs, vines, ferns and other herbaceous perennials form clonal colonies. Parts of a large clonal colony often become detached from the parent, termed fragmentation, to form separate individuals. Some plants also form seeds is asexually, termed apomixis, e.g. dandelion.

### Parthenogenesis

Clonal derivation exists in nature in some animal species and is referred to as parthenogenesis (reproduction of an organism by itself without a mate). This is an asexual form of reproduction that is only found in females of some insects, crustaceans and lizards. The growth and development occurs without fertilization by a male. In plants, parthenogenesis means the development of an embryo from an unfertilized egg cell, and is a component process of apomixis. In species that use the XY sex-determination system, the offspring will always be female. An example is the "Little Fire Ant" (*Wasmannia auropunctata*), which is native to Central and South America but has spread throughout many tropical environments.

## Artificial Cloning of Organisms

Artificial cloning of organisms may also be called *reproductive cloning.*

## Methods

Reproductive cloning generally uses "somatic cell nuclear transfer" (SCNT) to create animals that are genetically identical. This process entails the transfer of a nucleus from a donor adult cell (somatic cell) to an egg that has no nucleus. If the egg begins to divide normally it is transferred into the uterus of the surrogate mother. Such clones are not strictly identical since the somatic cells may contain mutations in their nuclear DNA. Additionally, the mitochondria in the cytoplasm also contains DNA and during SCNT this DNA is wholly from the donor egg, thus the mitochondrial genome is not the same as that of the nucleus donor cell from which it was produced. This may have important implications for cross-species nuclear transfer in which nuclear-mitochondrial incompatibilities may lead to death. Artificial *embryo splitting* or *embryo twinning* may also be used as a method of cloning, where an embryo is split in the maturation before embryo transfer. It is optimally performed at the 6- to 8-cell stage, where it can be used as an expansion of IVF to increase the number of available embryos. If both embryos are successful, it gives rise to monozygotic (identical) twins.

## Dolly (Sheep)

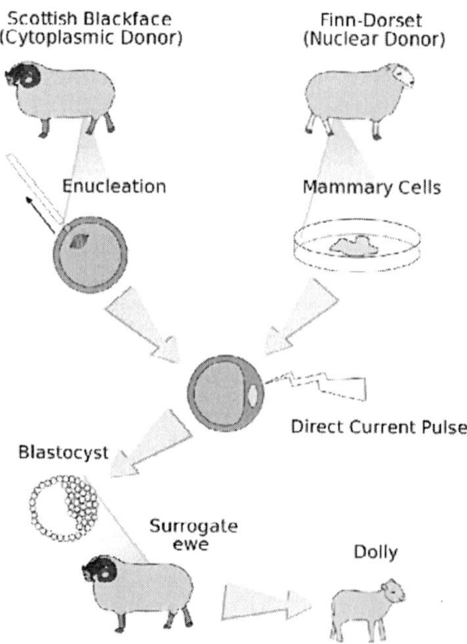


***Figure:*** *Dolly clone*

Dolly (5 July 1996 – 14 February 2003) was a female domestic sheep, and the first mammal to be cloned from an adult somatic cell, using the process of nuclear transfer. She was cloned by Ian Wilmut, Keith Campbell and colleagues at the Roslin Institute near Edinburgh in Scotland. She was born on 5 July 1996 and she lived until the age of six. She has been called "the world's most famous sheep" by sources including BBC News and *Scientific American.*

The cell used as the donor for the cloning of Dolly was taken from a mammary gland, and the production of a healthy clone therefore proved that a cell taken from a specific part of the body could recreate a whole individual. On Dolly's name, Wilmut stated "Dolly is derived from a mammary gland cell and we couldn't think of a more impressive pair of glands than Dolly Parton's"..

Dolly was born 5 July 1996 to three mothers (one provided the egg, another the DNA and a third carried the cloned embryo to term). She was created using the technique of somatic cell nuclear transfer, where the cell nucleus from an adult cell is transferred into an unfertilised oocyte (developing egg cell) that has had its nucleus removed. The hybrid cell is then stimulated to divide by an electric shock, and when it develops into a blastocyst it is implanted in a surrogate mother.

Dolly was the first clone produced from a cell taken from an adult mammal. The production of Dolly showed that genes in the nucleus of such a mature differentiated somatic cell are still capable of reverting back to an embryonic totipotent state, creating a cell that can then go on to develop into any part of an animal.

Dolly lived for her entire life at the Roslin Institute in Edinburgh. There she was bred with a Welsh Mountain ram and produced six lambs in total. Her first lamb, named Bonnie, was born in April 1998. The next year Dolly produced twin lambs Sally and Rosie, and she gave birth to triplets Lucy, Darcy and Cotton in the year after that. In the autumn of 2001, at the age of five, Dolly developed arthritis and began to walk stiffly, but this was successfully treated with anti-inflammatory drugs.

On 14 February 2003, Dolly was euthanised because she had a progressive lung disease and severe arthritis. A Finn Dorset such as Dolly has a life expectancy of around 11 to 12 years, but Dolly lived to be only six years of age. A post-mortem examination showed she had a form of lung cancer called Jaagsiekte, which is a fairly common disease of sheep and is caused by the retrovirus JSRV. Roslin scientists stated that they did not think there was a connection with Dolly being a clone, and that other sheep in the same flock had died of the same disease. Such lung diseases are a particular danger for sheep kept indoors, and Dolly had to sleep inside for security reasons.

Some have speculated that a contributing factor to Dolly's death was that she could have been born with a genetic age of six years, the same age as the sheep from which she was cloned. One basis for this idea was the finding that Dolly's telomeres were short, which typically is a result of the ageing process. The Roslin Institute have stated that intensive health screening did not reveal any abnormalities in Dolly that could have come from advanced ageing.

After cloning was successfully demonstrated through the production of Dolly, many other large mammals have been cloned, including horses and bulls. The attempt to clone argali (mountain sheep) did not produce viable embryos. The attempt to clone a banteng bull was more successful, as were the attempts to clone mouflon (a form of wild sheep), both resulting in viable offspring.

The reprogramming process cells need to go through during cloning is not perfect and embryos produced by nuclear transfer often show abnormal development making cloning mammals highly inefficient (Dolly was the only lamb that survived to adulthood from 277 attempts). Wilmut, who led the team that created Dolly, announced in 2007 that the nuclear transfer technique may never be sufficiently efficient for use in humans. Cloning may have uses in preserving endangered species and may become a viable tool for reviving extinct species. In January 2009, scientists from the Centre of Food Technology and Research of Aragon, in Zaragoza, northern Spain announced the cloning of the Pyrenean ibex, a form of wild mountain goat, which was officially declared extinct in 2000.

Although the newborn ibex died shortly after birth due to physical defects in its lungs it is the first time an extinct animal has been cloned, and may open doors for saving endangered and newly extinct species by resurrecting them from frozen tissue. Cloning does not introduce new genes into a population so will not increase genetic diversity. Cloning of domesticated animals could be important in the future production of transgenic livestock.

### *Species Cloned*

The modern cloning techniques involving nuclear transfer have been successfully performed on several species. Landmark experiments in chronological order:

In the acclaimed comic book Y: The Last Man the first human clone is believed to trigger a biological reaction leading to the death of almost all men on the planet.

Radiohead album Kid A has been suggested to be the story of the first human clone.

## *Religious Objections*

The Roman Catholic Church, under the papacy of Benedict XVI, has condemned the practice of human cloning, in the magisterial instruction *Dignitas Personae*, stating that it represents a "grave offence to the dignity of that person as well as to the fundamental equality of all people".

Sunni Muslims consider human cloning to be forbidden by Islam. The Islamic Fiqh Academy, in its Tenth Conference proceedings, which was convened in Jeddah, Saudi Arabia in the period from June 28, 1997 to July 3, 1997, issued a Fatwâ stating that human cloning is haraam (prohibited by the faith).

## *Cloning Extinct and Endangered Species*

Cloning, or more precisely, the reconstruction of functional DNA from extinct species has, for decades, been a dream of some scientists. The possible implications of this were dramatized in the best-selling novel by Michael Crichton and high budget Hollywood thriller *Jurassic Park*. In real life, one of the most anticipated targets for cloning was once the Woolly Mammoth, but attempts to extract DNA from frozen mammoths have been unsuccessful, though a joint Russo-Japanese team is currently working toward this goal.

In 2001, a cow named Bessie gave birth to a cloned Asian gaur, an endangered species, but the calf died after two days. In 2003, a banteng was successfully cloned, followed by three African wildcats from a thawed frozen embryo. These successes provided hope that similar techniques (using surrogate mothers of another species) might be used to clone extinct species. Anticipating this possibility, tissue samples from the last *bucardo* (Pyrenean Ibex) were frozen in liquid nitrogen immediately after it died in 2000. Researchers are also considering cloning endangered species such as the giant panda, ocelot, and cheetah. The "Frozen Zoo" at the San Diego Zoo now stores frozen tissue from the world's rarest and most endangered species.

In 2002, geneticists at the Australian Museum announced that they had replicated DNA of the Thylacine (Tasmanian Tiger), extinct about 65 years previous, using polymerase chain reaction. However, on February 15, 2005 the museum announced that it was stopping the project after tests showed the specimens' DNA had been too badly degraded by the (ethanol) preservative. On 15 May 2005 it was announced that the Thylacine project would be revived, with new participation from researchers in New South Wales and Victoria.

In January 2009, for the first time, an extinct animal, the Pyrenean ibex mentioned above was cloned, at the Centre of Food Technology and Research of Aragon, using the preserved DNA of the skin samples from 2001 and domestic goat egg-cells. (The ibex died shortly after birth due to physical defects in its lungs.) One of the continuing obstacles in the attempt to clone extinct species is the need for nearly perfect DNA. Cloning from a single specimen could not create a viable breeding population in sexually reproducing animals. Furthermore, even if males and females were to be cloned, the question would remain open whether they would be viable at all in the absence of parents that could teach or show them their natural Behaviour.

Cloning endangered species is a highly ideological issue. Many conservation biologists and environmentalists vehemently oppose cloning endangered species — mainly because they think it may deter donations to help preserve natural habitat and wild animal populations. The "rule-of-thumb" in animal conservation is that, if it is still feasible to conserve habitat and viable wild populations, breeding in captivity should not be undertaken in isolation.

In a 2006 review, David Ehrenfeld concluded that cloning in animal conservation is an experimental technology that, at its state in 2006, could not be expected to work except by pure chance and utterly failed a cost-benefit analysis. Furthermore, he said, it is likely to siphon funds from established and working projects and does not address any of the issues underlying animal extinction (such as habitat destruction, hunting or other overexploitation, and an impoverished gene pool). While cloning technologies are well-established and used on a regular basis in plant conservation, care must be taken to ensure genetic diversity. He concluded:

Vertebrate cloning poses little risk to the environment, but it can consume scarce conservation resources, and its chances of success in preserving species seem poor. To date, the conservation benefits of transgenics and vertebrate cloning remain entirely theoretical, but many of the risks are known and documented. Conservation biologists should devote their research and energies to the established methods of conservation, none of which require transgenics or vertebrate cloning.

## Genetically Modified Food

Genetically modified (GM) foods are foods derived from genetically modified organisms. Genetically modified organisms have had specific changes introduced into their DNA by genetic engineering techniques. These techniques are much more precise than mutagenesis (mutation breeding) where an organism is exposed to radiation or chemicals to create a non-specific but stable change. Other techniques by which humans modify food organisms include selective breeding (plant breeding and animal breeding), and somaclonal variation.

GM foods were first put on the market in the early 1990s. Typically, genetically modified foods are transgenic plant products: soybean, corn, canola, and cotton seed oil. Animal products have also been developed, although as of July 2010 none are currently on the market. In 2006 a pig was controversially engineered to produce omega-3 fatty acids through the expression of a roundworm gene. Researchers have also developed a genetically-modified breed of pigs that are able to absorb plant phosphorus more efficiently, and as a consequence the phosphorus content of their manure is reduced by as much as 60%.

Critics have objected to GM foods on several grounds, including possible safety issues, ecological concerns, and economic concerns raised by the fact that these organisms are subject to intellectual property law.

### Method

Genetic modification involves the insertion or deletion of genes. In the process of cisgenesis, genes are artificially transferred between organisms that could be conventionally bred. In the process of transgenesis, genes from a different species are inserted, which is a form of horizontal gene transfer. In nature this can occur when exogenous DNA penetrates the cell membrane for any reason. To do this artificially may require attaching genes to a virus or just physically inserting the extra DNA into the nucleus of the intended host with a very small syringe, or with very small particles fired from a gene gun. However, other methods exploit natural forms of gene transfer, such as the ability of Agrobacterium to transfer genetic material to plants, and the ability of lentiviruses to transfer genes to animal cells.

### Development

The first commercially grown genetically modified whole food crop was a tomato (called FlavrSavr), which was modified to ripen without softening, by Calgene, later a subsidiary of Monsanto. Calgene took the initiative to obtain FDA approval for its release in 1994 without any special labelling, although legally no such approval was required. It was welcomed by consumers who purchased the fruit at a substantial premium over the price of regular tomatoes.

However, production problems and competition from a conventionally bred, longer shelf-life variety prevented the product from becoming profitable. A tomato produced using similar technology to the Flavr Savr was used by Zeneca to produce tomato paste which was sold in Europe during the summer of 1996. The labelling and pricing were designed as a marketing experiment, which proved, at the time, that European consumers would accept genetically engineered foods. Currently, there are a number of food species in which a genetically modified version exists (percent modified are mostly 2009/2010 data).

### Growing GM Crops

Between 1997 and 2009, the total surface area of land cultivated with GMOs had increased by a factor of 80, from 17,000 km (4.2 million$^2$ acres) to 1,340,000 km$^2$ (331 million acres). Although most GM crops are grown in North America, in recent years there has been rapid growth in the area sown in developing countries. For instance in 2009 the largest increase in crop area planted to GM crops (soybeans) was in Brazil (214,000 km$^2$ in 2009 versus 158,000 km$^2$ in 2008.) There has also been rapid and continuing expansion of GM cotton varieties in India since 2002. (Cotton is a major source of vegetable cooking oil and animal feed.) In 2009 84,000 km$^2$ of GM cotton were harvested in India.

In India, GM cotton yields in Andhra Pradesh were no better than non-GM cotton in 2002, the first year of commercial GM cotton planting. This was because there was a severe drought in Andhra Pradesh that year and the parental cotton plant used in the genetic engineered variant was not well suited to extreme drought. Maharashtra, Karnataka, and Tamil Nadu had an average 42% increase in yield with GM cotton in the same year. Drought resistant variants were developed and, with the substantially reduced losses to insect predation, by 2009 87% of Indian cotton was GM. Though disputed the economic and environmental benefits of GM cotton in India to the individual farmer have been documented.

In 2009, countries that grew 95% of the global transgenic crops were the United States (46%), Brazil (16%), Argentina (15%), India (6%), Canada (6%), China (3%), Paraguay (2%) and South Africa (2%).

the US that Bt-corn pollen might affect the monarch butterfly. However this concern was disproved by six comprehensive articles in the Proceedings of the National Academy of Sciences in 2001. Monarch populations increased, despite increased Bt corn probably due to reduced pesticide use. Other possible effects might come from the spread of genes from modified plants to unmodified relatives, which might produce species of weeds resistant to herbicides. In some areas of the US "superweeds" have evolved naturally, these weeds are resistant to herbicides and have forced farmers to return to traditional crop management practices.

There has been controversy over the results of a farm-scale trial in the United Kingdom comparing the impact of GM crops and conventional crops on farmland biodiversity. Some claimed that the results showed that GM crops had a significant negative impact on wildlife. They pointed out that the studies showed that using herbicide resistant GM crops allowed better weed control and that under such conditions there were fewer weeds and fewer weed seeds. This result was then extrapolated to suggest that GM crops would have significant impact on the wildlife that might rely on farm weeds. The President of the Royal Society, the body that had carried out the trials, stated that "To generalize and declare 'all GM is bad' or 'all GM is good' for the environment as a result of these experiments is a gross over-simplification", arguing that although the trials showed that the combination of some GM crops with long-lasting herbicides were bad for biodiversity, using other GM crops without these herbicides increased biodiversity.

In July 2005 British scientists showed that transfer of a herbicide-resistance gene from GM oilseed rape to a wild cousin, charlock, and wild turnips was possible.

Many agricultural scientists and food policy specialists view GM crops as an important element in sustainable food security and environmental management. This point of view is summarized in the ABIC Manifesto:

On our planet, 18% of the land mass is used for agricultural production. This fraction cannot be increased substantially. It is absolutely essential that the yield per unit of land increases beyond current levels given that: The human population is still growing, and will reach about nine billion by 2040;70,000 km$^2$ of agricultural land (equivalent to 60% of the German agricultural area) are lost annually to growth of cities and other non-agricultural uses; Consumer diets in developing countries are increasingly changing from plant-based proteins to animal protein, a trend that requires a greater amount of crop-based feeds.

Other scientists, such as Dr. Charles Benbrook, argue that improvement of global food security is hardly being addressed by genetic research and that a lack of yield is often not caused by insufficient genetic resources. Regarding the issues of intellectual property and patent law, an international report from the year 2000 states:

If the rights to these tools are strongly and universally enforced- and not extensively licensed or provided pro bono in the developing world- then the potential applications of GM technologies described previously are unlikely to benefit the less developed nations of the world for a long time (ie until after the restrictions conveyed by these rights have expired).

### Issues with Bt Maize

A well publicized claim associated with Bt crops (or transgenic maize) was the concern that pollen from Bt maize might kill the monarch butterfly. This report was puzzling because the pollen from most maize hybrids contains much lower levels of Bt than the rest of the plant and led to multiple follow-up studies. One possible issue revealed in these studies is the possibility that the initial study was flawed; based on the way the pollen was collected, in that they collected and fed non-toxic pollen that was mixed with anther walls that did contain Bt toxin. A collaborative research exercise was carried out over two years by several groups of scientists in the US and Canada, looking at the effects of Bt pollen in both the field and the laboratory. This resulted in a risk assessment that concluded that any risk posed by the corn to butterfly populations under real-world conditions was negligible. The USDA has stated that the weight of the evidence is that Bt crops do not pose a risk to the monarch butterfly. An independent 2002 review of the scientific literature concluded that "the commercial large-scale cultivation of current Bt–maize hybrids did not pose a significant risk to the monarch population" and noted that despite large-scale planting of GM crops, the butterfly's population is increasing.

In 2007 Andreas Lang, Éva Lauber and Bela Darvas criticized these studies, arguing that there can be a great difference in the effects between the acute exposure tested for and chronic exposure. Moreover, they stated that the "worst case conditions" performed were not in fact worst case scenarios, as laboratory conditions with ample food supply and a favourable climate ensure healthy subjects. They instead believe that in the wild, low temperatures, rain and parasites and disease might exacerbate a Bt effect on butterfly larvae.

Their own experiments suggested that some butterfly species were negatively affected by such chronic exposure. Jorg Romeis, who conducted the original studies, replied that if species of Butterfly are affected as Darvas claims that a "more comprehensive assessment will be needed and, depending on the degree and nature of concern, this may extend to field testing". A 2001 report in *Nature* presented evidence that Bt maize was cross-breeding with unmodified maize in Mexico, although the data in this paper was later described as originating from an artifact and *Nature* stated that "the evidence available is not sufficient to justify the publication of the original paper". A subsequent large-scale study, in 2005, failed to find any evidence of contamination in Oaxaca. However, other authors have stated that they also found evidence of cross-breeding between natural maize and transgenic maize.

There is also a risk that for example, transgenic maize will crossbreed with wild grass variants, and that the Bt-gene will end up in a natural environment, retaining its toxicity. An event like this would have ecological implications. However, there is no evidence of crossbreeding between maize and wild grasses. In 2009 it was reported that 82,000 hectares (200,000 acres) of Bt corn in South Africa failed to produce seeds. Monsanto claimed average yield was reduced by 25% in those fields affected, it compensated the farmers concerned and the corn varieties were affected by a mistake made in the seed breeding process. Marian Mayet an environmental activitist and director of the Africa-centre for biosecurity in Johannesburg called for a government investigation and asserted that the biotechnology was at fault, "*You cannot make a 'mistake' with three different varieties of corn*". In 2009 South African farmers planted 1,900,000 hectares (4,700,000 acres) of GM maize (73% of the total crop).

As of 2007, a phenomenon called Colony Collapse Disorder (CCD) was noticed in bee hives all over North America, and elsewhere. Although it is not certain if this is a new phenomenon, initial ideas on the possible causes ranged from poor nutrition, infections, parasites and pesticide use. More unusual speculations included radio waves from cellphone base stations, climate change, and the use of transgenic crops containing Bt. The Mid-Atlantic Apiculture Research and Extension Consortium published a report on 2007-03-27 that found no evidence that pollen from Bt crops is adversely affecting bees. Several researchers in the US have since attributed CCD to the spread of a new virus called Israeli acute paralysis virus, although other parasites have also been implicated.

## Legal Issues in the US

### *Alfalfa*

On 21 June 2010, the US Supreme Court in *Monsanto v. Geertson Seed Farms* issued its first ruling about GM crops. This case relates to Roundup Ready alfalfa and dates back to 2006, when a coalition of groups led by the Centre for Food Safety raised concerns about environmental impacts that the United States Department of Agriculture (USDA) allegedly failed to address before approving the planting of Roundup Ready alfalfa. In response, in 2007, Judge Charles Breyer of the California Northern District Court ruled that USDA was in error when it approved the planting of Roundup Ready alfalfa.

In June 2009, a divided three-judge panel on the 9th U.S. Circuit Court of Appeals upheld Judge Charles Breyer's decision, and Monsanto appealed to the Supreme Court.

The impact of the current US Supreme Court ruling is somewhat unclear, with both sides appearing to claim victory.

### *Sugar Beets*

There is also a somewhat similar case involving sugar beets before the same California Northern District Court. This is a case involving Monsanto's breed of pesticide-resistant sugar beets. While Judge Jeffrey S. White (issuing his ruling in the spring of 2010) allowed the planting of GM sugar beets to continue, he also warned that this may be blocked in the future while an environmental review is taking place.

Finally, on 13 August 2010, Judge Jeffrey S. White ordered the halt to the planting of the genetically modified sugar beets in the US. He indicated that "the Agriculture Department had not adequately assessed the environmental consequences before approving them for commercial cultivation."

In 2010, before the ruling, 95% of the sugar beet grown in the US was GM. About half the sugar supply in the US came from sugar beet.

### *Control of the Market*

Patent holder companies use their control of their own GMO to corner the market and gain profit. However, other companies often compete for the little market share available to GM foods worldwide. Detractors such as Greenpeace say that patent rights give corporations a dangerous amount of control over their product while corporations

- Tadpole: (1952) Many scientists questioned whether cloning had actually occurred and unpublished experiments by other labs were not able to reproduce the reported results.
- Carp: (1963) In China, embryologist Tong Dizhou produced the world's first cloned fish by inserting the DNA from a cell of a male carp into an egg from a female carp. He published the findings in a Chinese science journal.
- Mice: (1986) A mouse was the first mammal successfully cloned from an early embryonic cell. Soviet scientists Chaylakhyan, Veprencev, Sviridova, and Nikitin had the mouse "Masha" cloned. Research was published in the magazine "Biofizika" volume II, issue 5 of 1987.
- Sheep: (1996) From early embryonic cells by Steen Willadsen. Megan and Morag cloned from differentiated embryonic cells in June 1995 and Dolly the sheep from a somatic cell in 1997.
- Rhesus Monkey: Tetra (January 2000) from embryo splitting
- Gaur: (2001) was the first endangered species cloned.
- Cattle: Alpha and Beta (males, 2001) and (2005) Brazil
- Cat: CopyCat "CC" (female, late 2001), Little Nicky, 2004, was the first cat cloned for commercial reasons
- Dog: Snuppy, a male Afghan hound was the first cloned dog (2005).
- Rat: Ralph, the first cloned rat (2003)
- Mule: Idaho Gem, a john mule born 4 May 2003, was the first horse-family clone.
- Horse: Prometea, a Haflinger female born 28 May 2003, was the first horse clone.
- Water Buffalo: Samrupa was the first cloned water buffalo. It was born on February 6, 2009, at India's Karnal National Diary Research Institute but died five days later due to lung infection.
- Camel: (2009) Injaz, is the first cloned camel.

## Human Cloning

Human cloning is the creation of a genetically identical copy of a human. It does not usually refer to monozygotic multiple births, human cell or tissue reproduction. The ethics of cloning is an extremely controversial issue. The term is generally used to refer to artificial human cloning; human clones in the form of identical twins are commonplace, with their cloning occurring during the natural process of reproduction. There are two commonly discussed types of human cloning: *therapeutic cloning* and *reproductive cloning*. Therapeutic cloning involves cloning cells from an adult for use in medicine and is an active area of research, while reproductive cloning would involve making cloned humans. Such reproductive cloning has not been performed and is illegal in many countries.

A third type of cloning called replacement cloning is a theoretical possibility, and would be a combination of therapeutic and reproductive cloning. Replacement cloning would entail the replacement of an extensively damaged, failed, or failing body through cloning followed by whole or partial brain transplant.

## History

Although the possibility of cloning humans has been the subject of speculation for much of the twentieth century, scientists and policy makers began to take the prospect seriously in the 1960s. Nobel Prize winning geneticist Joshua Lederberg advocated for cloning and genetic engineering in a seminal article in the *American Naturalist* in 1966 and again, the following year, in the *Washington Post*. He sparked a debate with conservative bioethicist Leon Kass, who wrote at the time that "the programmed reproduction of man will, in fact, dehumanize him." Another Nobel Laureate, James D. Watson, publicized the potential and the perils of cloning in his *Atlantic Monthly* essay, "Moving Toward the Clonal Man", in 1971.

The technology of cloning mammals, although far from reliable, has reached the point where many scientists are knowledgeable, the literature is readily available, and the implementation of the technology is not very expensive compared to many other scientific processes. For that reason Lewis D. Eigen has argued that human cloning attempts will be made in the next few years and may well have been already begun. The ethical and moral issues cannot wait and should be discussed, debated and guidelines and laws be developed now.

*"By waiting until the first clone is among us or about to be born, we complicate the problem immensely and guarantee that we will not be able to have the national and international conversation and debate to arrive at particularly good decisions like using protection."*

## Notable Cloning Attempts and Claims

- Dr. Panayiotis Zavos, an American fertility doctor, revealed on 17 January 2004 at a London press conference that he had transferred a freshly-cloned embryo into a 35-year-old woman.

On 4 February 2004, it emerged that the attempt had not worked and the woman did not become pregnant.

## Ethics of Cloning

In bioethics, the ethics of cloning refers to a variety of ethical positions regarding the practice and possibilities of cloning, especially human cloning. While many of these views are religious in origin, the questions raised by cloning are faced by secular perspectives as well.

As the science of cloning continues to advance, governments have dealt with ethical questions through legislation.

### *Philosophical Debate*

Cloning, particularly human cloning, is highly controversial.

Advocates of human therapeutic cloning believe the practice could provide genetically identical cells for regenerative medicine, and tissues and organs for transplantation. Such cells, tissues, and organs would neither trigger an immune response nor require the use of immunosuppressive drugs. Both basic research and therapeutic development for serious diseases such as cancer, heart disease, and diabetes, as well as improvements in burn treatment and reconstructive and cosmetic surgery, are areas that might benefit from such new technology. One bioethicist, Jacob M. Appel of New York University, has gone so far as to argue that "children cloned for therapeutic purposes" such as "to donate bone marrow to a sibling with leukemia" may someday be viewed as heroes.

Proponents claim that human reproductive cloning also would produce benefits. Severino Antinori and Panos Zavos hope to create a fertility treatment that allows parents who are both infertile to have children with at least some of their DNA in their offspring.

Some scientists, including Dr. Richard Seed, suggest that human cloning might obviate the human aging process. How this might work is not entirely clear since the brain or identity would have to be transferred to a cloned body. Dr. Preston Estep has suggested the terms "replacement cloning" to describe the generation of a clone of a previously living person, and "persistence cloning" to describe SENS (Strategies for Engineered Negligible Senescence) one of the considered options to repair the cell depletion related to cellular senescence is to grow replacement tissues from stem cells harvested from a cloned embryo.

At present, the main non-religious objection to human cloning is that cloned individuals are often biologically damaged, due to the inherent unreliability of their origin; for example, researchers currently are unable to safely and reliably clone non-human primates. For example, bioethicist Thomas Murray of the Hastings Centre argues that "it is absolutely inevitable that groups are going to try to clone a human being. But they are going to create a lot of dead and dying babies along the way."

UNESCO's Universal Declaration on Human Genome and Human Rights asserts that cloning contradicts human nature and dignity: Cloning is an asexual reproductive mode, which could distort generation lines and family relationships, and limit genetic differentiation, which ensures that human life is largely unique. Cloning can also imply an instrumental attitude toward humans, which risks turning them into manufactured objects, and interferes with evolution, the implications of which we lack the insight or prescience to predict. Furthermore, proponents of animal rights argue that non-human animals possess certain moral rights as living entities and should therefore be afforded the same ethical considerations as human beings. This would negate the exploitation of animals in scientific research on cloning, cloning used in food production, or as other resources for human use or consumption.

Rudolph Jaenisch, a professor at Harvard, has pointed out that we have become more efficient at producing clones which are still defective. Other arguments against cloning come from various religious orders (believing cloning violates God's will or the natural order of life), and a general discomfort some have with the idea of "meddling" with the creation and basic function of life. This unease often manifests itself in contemporary novels, movies, and popular culture, as it did with numerous prior scientific discoveries and inventions. Various fictional scenarios portray clones being unhappy, soulless, or unable to integrate into society. Furthermore, clones are often depicted not as unique individuals but as "spare parts," providing organs for the clone's original (or any non-clone that requires replacement organs).

### Religious Views

***Christian:*** Roman Catholicism and many conservative Christian groups have opposed human cloning and the cloning of human embryos, since they believe that life begins at the moment of conception. Other Christian denominations such as the United Church of Christ do not believe a fertilized egg constitutes a living being, but still they oppose the cloning of embryonic cells. The World Council of Churches, representing nearly 400 Christian denominations worldwide, opposed cloning of both human embryos and whole humans in February 2006.

claim that they need product control in order to prevent seed piracy, fulfil financial obligations to shareholders and invest in further GM development.

Governments have also sought to protect their commercial interests through punitive measures against countries resisting GM foods on ethical or scientific grounds: after moves in France to ban a Monsanto GM corn variety, the US embassy recommended that 'we calibrate a target retaliation list that causes some pain across the EU'.

## Controversial Cases

### *Pusztai Potato Controversy*

A large media event occurred in 1998 when scientist Arpad Pusztai, who "was considered a world expert on plant lectins", reported in a television interview that he had found that rats fed potatoes genetically modified by the English biotech company *Cambridge Agricultural Genetics* to contain lectin, a natural insecticide in snowdrop plants, had stunted growth and immune system damage. Initially the Rowett Institute's director, Philip James, congratulated Pusztai after the interview, but his attitude changed after confusion arose over whether the lectin was from a jackbean or a snowdrop (the jackbean lectin is poisonous to the intestines).

A press release from the Rowett Institute stated that the lectin had come from a jackbean, but no experiments were made using lectins from the jackbean; in the confusion, it was alleged that Pusztai hadn't actually carried out the experiments. James suspended Pusztai and used misconduct procedures to seize the data. His annual contract was not renewed and both Pusztai and his wife were banned from speaking publicly. In October 1998 the Rowett Institute published an audit criticizing Pusztai's results, which, along with Pusztai's raw data, was sent to six anonymous reviewers who criticized Pusztai's work. Pusztai responded that the raw data was "never intended for publication under intense scrutiny". Pusztai sent the audit report and his rebuttal to scientists who requested it, and in February 1999, twenty-one European and American scientists, mostly friends or acquaintances of Pusztai, orchestrated by the environmental group Friends of the Earth, released a memo supporting Pusztai. Stanley Ewen, who worked with Pusztai, conducted a followup study supporting Pusztai's work and presented the work to a lectin meeting in Sweden.

In October 1999 Pusztai's research was published in the journal The Lancet. Because of the controversial nature of his research the data in this paper, co-authored by Stanley Ewen, was seen by a total of six reviewers when presented for peer review; five of these reviewers judged the work acceptable, although one of the five "deemed the study flawed but favoured publication to avoid suspicions of a conspiracy against Pusztai and to give colleagues a chance to see the data for themselves". The paper did not mention stunted growth or immunity issues, but reported that rats fed on potatoes genetically modified with the snowdrop lectin had "thickening in the mucosal lining of their colon and their jejunum" when compared with rats fed on non modified potatoes. Three Dutch scientists criticized the study on the grounds that the unmodified potatoes were not a fair control diet, and that any rats fed only on potatoes will suffer from a protein deficiency; Pusztai responded to these criticisms by stating that the protein and energy were comparable, and that "a sample size of six is perfectly normal in studies like this".

The genetically modified potatoes were later destroyed by *Cambridge Agricultural Genetics* as were all details of their modification.

### Bioequivalence Study of a Corn Cultivar

A controversy arose around biotech company Monsanto's data on a 90-Day Rat Feeding Study on the MON863 strain of GM corn. In May 2005, critics of GM foods pointed to differences in kidney size and blood composition found in this study, suggesting that the observed differences raises questions about the regulatory concept of substantial equivalence.

Anti-GM campaigner Jeffrey M. Smith, writing in Biophile Magazine, quoting comments from Pusztai and Seralini, has stated that nutritional studies typically use young, fast-growing animals with starting weights not varying by more than 2% from the average whereas Monsanto's research design used a mix of young and old animals with starting weights ranging from 198.4 to 259.8 grams. Seralini and two other authors published a study of these data, funded by Greenpeace, in 2007 making similar points.

The raising of this issue prompted the European Food Safety Authority (EFSA) to reexamine the safety data on this strain of corn. The EFSA concluded that the observed small numerical decrease in rat kidney weights were not biologically meaningful, and the weights were well within the normal range of kidney weights for control animals. There were no corresponding microscopic findings in the relevant organ systems, and they stated that all blood chemistry and organ weight values fell within the "normal range of historical control values" for rats. In addition the EFSA review stated that the statistical methods used by Seralini et al. in the analysis of the data were incorrect. The European Commission has approved the ÌÏÍ863 corn for animal and human consumption. Food Standards Australia New Zealand reviewed the 2007 Seralini et al study and concluded that "...all of the statistical differences between rats fed MON 863 corn and control rats are attributable to normal biological variation."

Greenpeace stated in a 2007 press release that Seralini et al. had completed a similar analysis of the NK603 strain of corn and came to similar conclusions as they did in their previous study. Seralini et al included this in a re-analysis of three existing rat feeding studies published in 2009.

The European Food Safety Authority reviewed the 2009 Seralini et al paper and concluded that the author's claims were not supported by the data in their paper. They noted that many of their fundamental statistical chriticisms of the 2007 paper also applied to the 2009 paper. There was no new information that would change the EFSA's conclusions that the three GM maize types were safe for human, animal health and the environment The French High Council of Biotechnologies Scientific Committee (HCB) also reviewed the 2009 study and concluded that it "..presents no admissible scientific element likely to ascribe any haematological, hepatic or renal toxicity to the three re-analysed GMOs." The HCB also questioned the author's independence. Food Standards Australia New Zealand concluded that the results from the 2009 Seralini et al. study were due to chance alone.

## Contamination Issues

In the 1990s genetically modified Flax tolerant to herbicide residues in soil was developed by the Crop Development Centre (CDC) at the University of Saskatchewan in Canada. Named Flax variety FP967, but commonly called *CDC Triffid*, research was controversially halted following protests from Canadian farmers who stood to lose up to 70% of their traditional export markets if it was introduced. GM Flax was deregistered, its sale was criminalized and in 2001 all modified seeds were destroyed. No modified crops had been planted and no seed had been sold but GM industry proponent Alan McHughen controversially passed out sample packets of seeds at presentations. In early September 2009, Flax imported into Germany was found to be contaminated with *CDC Triffid* causing the price of Canadian Flax to fall 32 percent. By mid November 35 countries reported contamination of imported

Canadian Flax which has now been banned by the European Union. Canadian farmers are expected to be responsible for the cost of the cleanup and testing of future crops.

In 2000, Aventis StarLink corn, which had been approved only as animal feed due to concerns about possible allergic reactions in humans, was found contaminating corn products in U.S. supermarkets. An episode involving Taco Bell taco shells was particularly well publicized which resulted in sales of StarLink seed being discontinued. The registration for the Starlink varieties was voluntarily withdrawn by Aventis in October 2000. Aid sent by the UN and the US to Central African nations was also found to be contaminated with StarLink corn and the aid was rejected. The US corn supply has been monitored for Starlink Bt proteins since 2001 and no positive samples have been found since 2004.

### Public Perception

Research by the Pew Initiative on Food and Biotechnology has shown that in 2005 Americans' knowledge of genetically modified foods and animals continues to remain low, and their opinions reflect that they are particularly uncomfortable with animal cloning. In one instance of consumer confusion, DNA Plant Technology's Fish tomato transgenic organism was conflated with Calgene's Flavr Savr transgenic food product. The Pew survey also showed that despite continuing concerns about GM foods, American consumers do not support banning new uses of the technology, but rather seek an active role from regulators to ensure that new products are safe.

Only 2% of Britons were said to be "happy to eat GM foods", and more than half of Britons were against GM foods being available to the public, according to a 2003 study. However a 2009 review article of European consumer polls concluded that opposition to GMOs in Europe has been gradually decreasing. Approximately half of European consumers accepted gene technology, particularly when benefits for consumers and for the environment could be linked to GMO products. 80 % of respondents did not cite the application of GMOs in agriculture as a significant environmental problem. Many consumers seem unafraid of health risks from GMO products and most European consumers did not actively avoid GMO products while shopping. In Australia, GM foods that have novel DNA, novel protein, altered characteristics or has to be cooked or prepared in a different way compared to the conventional food have, since December 2001, had to be identified on food labels.

However, multiple surveys have shown that while 45% of the public will accept GM foods, some 93% demand all genetically modified foods be labelled as such. A 2007 survey by the Food Standards Australia and New Zealand found that 27% of Australians looked at the label to see if it contained GM material when purchasing a grocery product for the first time.

Labelling legislation has been introduced and rejected several times since 1996 on the grounds of "restraint of trade" due to the cost of labelling. The controversy erupted again in 2009 when Graincorp, the nations largest grain handler, announced it would mix GM Canola with its unmodified grain. Traditional growers, who largely rely on GM-free markets, had been told they would need to pay to have their produce certified GM free. Graincorp reversed its decision the same year. Critics such as Greenpeace and the Gene Ethics Network have renewed calls for more labelling.

Opponents of genetically modified food often refer to it as "Frankenfood", after Mary Shelley's character Frankenstein and the monster he creates, in her novel of the same name. The term was coined in 1992 by Paul Lewis, an English professor at Boston College who used the word in a letter he wrote to the *New York Times* in response to the decision of the US Food and Drug Administration to allow companies to market genetically modified food. The term "Frankenfood" has become a battle cry of the European side in the US-EU agricultural trade war.

Critics have protested in regards to the appointment of pro GM lobbyists to senior positions in the FDA. Michael R. Taylor has been appointed as a senior adviser to the FDA on food safety and Dennis Wolff is expected to take up the position of Under-Secretary of the newly created Agriculture for Food Safety.

Taylor is a former Monsanto lobbyist credited as being responsible for the implementation of "substantial equivalence" in place of food safety studies and for his advocacy that resulted in the Delaney clause that prohibited the inclusion of "any chemical additive found to induce cancer in man.. or animals" in processed foods being amended in 1996 to allow the inclusion of pesticides in GMOs. Wolff is the Pennsylvania Secretary of Agriculture who successfully lobbied to ban organic farmers from labelling their products as being GM free and was a proponent of the "ACRE" initiative which gave the Pennsylvania state attorney general's office the authority to sue municipalities that banned GMOs. Several anti-GMO organisations have organised petitions demanding

Taylor's resignation and opposing Wolff's appointment and also conducted letter writing campaigns protesting the conflict of interest.

### *Religious Issues*

As of yet, no GM foods have been designated as unacceptable by religious authorities.

Economic and environmental effects:

- Many proponents of genetically engineered crops claim they lower pesticide usage and have brought higher yields and profitability to many farmers, including those in developing nations. For example, a 2010 study by US scientists, found that the economic benefit of Bt corn to farmers in five mid-west states was $6.9 billion over the previous 14 years. They were surprised that the majority ($4.3 billion) of the benefit accrued to non-Bt corn. This was speculated to be because the European Corn Borers that attack the Bt corn die and there are fewer left to attack the non-GM corn nearby.
- The United States has seen a widespread adoption of genetically-engineered corn, cotton and soybean crops since 1996.
- In 2010, the U.S. National Academy of Sciences reported that genetically engineered crops had resulted in reduced pesticide application and reduced soil erosion from tilling. The report also stated that the advent of glyphosate-herbicide resistant weeds—that have developed because of the use of engineered crops—could cause the genetically engineered crops to lose their effectiveness unless farmers also use other established weed management strategies.
- In a study by Scientists at the University of Arkansas published in 2010 showed that about 83 percent of wild or weedy canola they tested contained genetically modified herbicide resistance genes, and they also found some plants that contained resistance to both herbicides, a combination of transgenic traits that had not been developed in canola crops. That leads us to believe that these wild populations that contain modified genes have become established populations.

### *Bans*

- In 2002, Zambia cut off the flow of Genetically Modified Food (mostly maize) from UN's World Food Programme. This left a famine-stricken population without food aid.

- In December 2005 the Zambian government changed its mind in the face of further famine and allowed the importation of GM maize. However, the Zambian Minister for Agriculture Mundia Sikatana has insisted that the ban on genetically modified maize remains, saying "We do not want GM (genetically modified) foods and our hope is that all of us can continue to produce non-GM foods."
- In April 2004 Hugo Chavez announced a total ban on genetically modified seeds in Venezuela.
- In January 2005, the Hungarian government announced a ban on importing and planting of genetic modified maize seeds, which was subsequently authorized by the EU.
- On August 18, 2006, American exports of rice to Europe were interrupted when much of the U.S. crop was confirmed to be contaminated with unapproved engineered genes, possibly caused by cross-pollination with conventional crops.
- On February 9, 2010, Indian Environment Minister, Jairam Ramesh, imposed a moratorium on the cultivation of GMF "for as long as it is needed to establish public trust and confidence". His decision was made after protest from several groups responding to regulatory approval of the cultivation of Bt brinjal, a GM eggplant in October, 2009.

### *U.S. Government Reaction to European Ban*

In recent years, France and several other European countries banned Monsanto's MON-810 corn and similar genetically modified food crops. In late 2007, the U.S. ambassador to France recommended "moving to retaliation" against France and the European Union in an attempt to fight the French ban and changes in European policy toward genetically modified crops, according to a U.S. government diplomatic cable obtained by WikiLeaks. The U.S. ambassador to France recommended retaliation to cause "some pain across the EU."

### *Intellectual Property*

Traditionally, farmers in all nations saved their own seed from year to year. It should be noted that this does not apply in more agriculturally developed countries for some crops. Corn is one example where producers generally have not saved seed since the early 1900s with the advent of hybrid corn through selective breeding. Seed producers grow the seed corn instead due to the effort needed to produce hybrids. The offspring of the hybrid corn, while still viable, lose the beneficial traits of the parents, resulting in the loss of hybrid vigor. In these cases, the use of hybrid plants has been the primary reason for growers not saving seed, not intellectual property issues, and has been in practice well before genetically-modified seed was developed. However, the practice of not saving seed in non-hybrid crops, such as soybean, is mainly due to intellectual property regulations. Allowing to follow this practice with genetically modified seed would result in seed developers losing the ability to profit from their breeding work. Therefore, genetically-modified seed is subject to licensing by their developers in contracts that are written to prevent farmers from following this practice.

Enforcement of patents on genetically modified plants is often contentious, especially because of gene flow. In 1998, 95-98 percent of about 10 km planted with canola by Canadian farmer Percy Schmeiser were found to contain Monsanto Company's patented Roundup Ready gene although Schmeiser had never purchased seed from Monsanto. The initial source of the plants was undetermined, and could have been through either gene flow or intentional theft. However, the overwhelming predominance of the trait implied that Schmeiser must have intentionally selected for it. The court determined that Schmeiser had saved seed from areas on and adjacent to his property where Roundup had been sprayed, such as ditches and near power poles.

Although unable to prove direct theft, Monsanto sued Schmeiser for piracy since he knowingly grew *Roundup Ready* plants without paying royalties (Ibid). The case made it to the Canadian Supreme Court, which in 2004 ruled 5 to 4 in Monsanto's favour. The dissenting judges focused primarily on the fact that Monsanto's patents covered only the gene itself and glyphosate resistant *cells*, and failed to cover transgenic plants in their entirety. All of the judges agreed that Schmeiser would not have to pay any damages since he had not benefited from his use of the genetically modified seed.

In response to criticism, Monsanto Canada's Director of Public Affairs stated that "It is not, nor has it ever been Monsanto Canada's policy to enforce its patent on Roundup Ready crops when they are present on a farmer's field by accident...Only when there has been a knowing and deliberate violation of its patent rights will Monsanto act."

### *Future Developments*

Future envisaged applications of GMOs are diverse and include drugs in food, bananas that produce human vaccines against infectious diseases such as Hepatitis B, metabolically engineered fish that mature more quickly, fruit and nut trees that yield years earlier, foods no longer containing properties associated with common intolerances, and plants that produce new plastics with unique properties. While their practicality or efficacy in commercial production has yet to be fully tested, the next decade may see exponential increases in GM product development as researchers gain increasing access to genomic resources that are applicable to organisms beyond the scope of individual projects. Safety testing of these products will also, at the same time, be necessary to ensure that the perceived benefits will indeed outweigh the perceived and hidden costs of development. Plant scientists, backed by results of modern comprehensive profiling of crop composition, point out that crops modified using GM techniques are less likely to have unintended changes than are conventionally bred crops.

### Health Risks

In the United States, the FDA Centre for Food Safety and Applied Nutrition reviews summaries of food safety data developed and voluntarily submitted by developers of engineered foods, in part on the basis of comparability to conventionally-produced foods. There are no specific tests required by FDA to determine safety. FDA does not approve the safety of engineered foods, but after its review, acknowledges that the developer of the food has asserted that it is safe.

A 2008 review published by the Royal Society of Medicine noted that GM foods have been eaten by millions of people worldwide for over 15 years, with no reports of ill effects. Similarly a 2004 report from the US National Academies of Sciences stated: "To date, no adverse health effects attributed to genetic engineering have been documented in the human population." There have, however, been no epidemiological studies to determine whether engineered crops have caused any harm to the public. Without such studies, it is unlikely that harm, if it occurred, would be detected or attributed to engineered foods. Worldwide, there are a range of perspectives within non-governmental organizations on the safety of GM foods. For example, the US pro-GM pressure group AgBioWorld has argued that GM foods have been proven safe, while other pressure groups and consumer rights groups, such as the Organic Consumers Association, and Greenpeace claim the long term health risks which GM could pose, or the environmental risks associated with GM, have not yet been adequately investigated.

In 1998 Rowett Research Institute scientist Arpad Pusztai reported that consumption of potatoes genetically modified to contain lectin had adverse intestinal effects on rats. Pusztai eventually published a paper, co-authored by Stanley Ewen, in the journal, The Lancet. The paper claimed to show that rats fed on potatoes genetically modified with the snowdrop lectin had unusual changes to their gut tissue when compared with rats fed on non modified potatoes.

The experiment modified potatoes to add a toxin (snowdrop lectin), but the experiment failed to include a control for the toxin alone or a control for genetic modifications alone (without added toxin); therefore, no conclusion could be made about the safety of the genetic engineering. The experiment has been criticised by other scientists on the grounds that the unmodified potatoes were not a fair control diet and that all the rats may have been sick, due to them being fed a diet of only potatoes.

In 2009 three scientists (Vendomois et al) published a statistical re-analysis of three feeding trials that had previously been published by others as establishing the safety of genetically modified corn. The new article claimed that their statistics instead showed that the three patented crops (Mon 810, Mon 863, and NK 603) developed and owned by Monsanto cause liver, kidney, and heart damage in mammals. A 2007 analysis of part of this data by the same group of scientists funded by Greenpeace was assessed by a panel of independent toxicologists in a study funded by Monsanto and published in the journal *Food and chemical toxicology.*

The reviewers reported that the study was statistically flawed and providing no evidence of adverse effects. The French High Council of Biotechnologies Scientific Committee reviewed the 2009 Vendomois et al study and concluded that it "..presents no admissible scientific element likely to ascribe any haematological, hepatic or renal toxicity to the three re-analysed GMOs." An evaluation by the European Food Safety Authority of the 2009 and 2007 studies noted that most of the results were within natural variation and they did not consider any of the effects reported biologically relevant. A review by Food Standards Australia New Zealand of the 2009 Vendomois et al study concluded that the results were due to chance alone.

### Gene Transfer

As of January 2009 there has only been one human feeding study conducted on the effects of genetically modified foods. The study involved seven human volunteers who had previously had their large intestines removed. These volunteers were to eat GM soy to see if the

DNA of the GM soy transferred to the bacteria that naturally lives in the human gut. Researchers identified that three of the seven volunteers had transgenes from GM soya transferred into the bacteria living in their gut before the start of the feeding experiment. As this low-frequency transfer did not increase after the consumption of GM Soy, the researchers concluded that gene transfer did not occur during the experiment. In volunteers with complete digestive tracts, the transgene did not survive passage through intact gastrointestinal tract. Anti-GM advocates believe the study should prompt additional testing to determine its significance.. Other studies have found DNA from M13 virus, GFP and even ribulose-1,5-bisphosphate carboxylase (Rubisco) genes in the blood and tissue of ingesting animals.

Two studies on the possible effects of feeding genetically modified feeds to animals found that there was no significant differences in the safety and nutritional value of feedstuffs containing material derived from genetically modified plants. Specifically, the studies noted that no residues of recombinant DNA or novel proteins have been found in any organ or tissue samples obtained from animals fed with GMP plants.

### *Allergies*

In the mid 1990s Pioneer Hi-Bred tested the allergenicity of a transgenic soybean that expressed a Brazil nut seed storage protein in hope that the seeds would have increased levels of the amino acid methionine. The tests (radioallergosorbent testing, immunoblotting, and skin-prick testing) showed that individuals allergic to Brazil nuts were also allergic to the new GM soybean. Pioneer has indicated that it will not develop commercial cultivars containing Brazil nut protein because the protein is likely to be an allergen.

### *Traceability*

In a January 2010 paper by Costa et al. the extraction and detection of DNA along a complete industrial soybean oil processing chain was described to monitor the presence of Roundup Ready (RR) soybean: "The amplification of soybean lectin gene by end-point polymerase chain reaction (PCR) was successfully achieved in all the steps of extraction and refining processes, until the fully refined soybean oil. The amplification of RR soybean by PCR assays using event-specific primers was also achieved for all the extraction and refining steps, except for the intermediate steps of refining (neutralisation, washing and bleaching) possibly due to sample instability. The real-time PCR assays using specific probes confirmed all the results and proved that it is possible to detect and quantify genetically modified organisms in the fully refined soybean oil. To our knowledge, this has never been reported before and represents an important accomplishment regarding the traceability of genetically modified organisms in refined oils."

### *Reduced Vulnerability of Crops to Environmental Stresses*

Crops containing genes that will enable them to withstand biotic and abiotic stresses may be developed. For example, drought and excessively salty soil are two important limiting factors in crop productivity. Biotechnologists are studying plants that can cope with these extreme conditions in the hope of finding the genes that enable them to do so and eventually transferring these genes to the more desirable crops.

One of the latest developments is the identification of a plant gene, At-DBF2, from Arabidopsis thaliana, a tiny weed that is often used for plant research because it is very easy to grow and its genetic code is well mapped out. When this gene was inserted into tomato and tobacco cells, the cells were able to withstand environmental stresses like salt, drought, cold and heat, far more than ordinary cells. If these preliminary results prove successful in larger trials, then At-DBF2 genes can help in engineering crops that can better withstand harsh environments. Researchers have also created transgenic rice plants that are resistant to rice yellow mottle virus (RYMV). In Africa, this virus destroys majority of the rice crops and makes the surviving plants more susceptible to fungal infections.

### *Increased Nutritional Qualities*

Proteins in foods may be modified to increase their nutritional qualities. Proteins in legumes and cereals may be transformed to provide the amino acids needed by human beings for a balanced diet. A good example is the work of Professors Ingo Potrykus and Peter Beyer in creating Golden rice.

### *Improved Taste, Texture or Appearance of Food*

Modern biotechnology can be used to slow down the process of spoilage so that fruit can ripen longer on the plant and then be transported to the consumer with a still reasonable shelf life. This alters the taste, texture and appearance of the fruit. More importantly, it could expand the market for farmers in developing countries due to the reduction in spoilage. However, there is sometimes a lack of understanding by researchers in developed countries about the actual needs of prospective beneficiaries in developing countries. For example, engineering soybeans to resist spoilage makes them less suitable for producing tempeh which is a significant source of protein that depends on fermentation. The use of modified soybeans results in a lumpy texture that is less palatable and less convenient when cooking.

The first genetically modified food product was a tomato which was transformed to delay its ripening. Researchers in Indonesia, Malaysia, Thailand, Philippines and Vietnam are currently working on delayed-ripening papaya in collaboration with the University of Nottingham and Zeneca.

Biotechnology in cheese production: enzymes produced by micro-organisms provide an alternative to animal rennet – a cheese coagulant – and an alternative supply for cheese makers. This also eliminates possible public concerns with animal-derived material, although there are currently no plans to develop synthetic milk, thus making this argument less compelling. Enzymes offer an animal-friendly alternative to animal rennet. While providing comparable quality, they are theoretically also less expensive. About 85 million tons of wheat flour is used every year to bake bread. By adding an enzyme called maltogenic amylase to the flour, bread stays fresher longer. Assuming that 10–15% of bread is thrown away as stale, if it could be made to stay fresh another 5–7 days then perhaps 2 million tons of flour per year would be saved. Other enzymes can cause bread to expand to make a lighter loaf, or alter the loaf in a range of ways.

### Reduced Dependence on Fertilizers, Pesticides and other Agrochemicals

Most of the current commercial applications of modern biotechnology in agriculture are on reducing the dependence of farmers on agrochemicals. For example, *Bacillus thuringiensis* (Bt) is a soil bacterium that produces a protein with insecticidal qualities. Traditionally, a fermentation process has been used to produce an insecticidal spray from these bacteria. In this form, the Bt toxin occurs as an inactive protoxin, which requires digestion by an insect to be effective. There are several Bt toxins and each one is specific to certain target insects.

Crop plants have now been engineered to contain and express the genes for Bt toxin, which they produce in its active form. When a susceptible insect ingests the transgenic crop cultivar expressing the Bt protein, it stops feeding and soon thereafter dies as a result of the Bt toxin binding to its gut wall. Bt corn is now commercially available in a number of countries to control corn borer (a lepidopteran insect), which is otherwise controlled by spraying (a more difficult process).

Crops have also been genetically engineered to acquire tolerance to broad-spectrum herbicide. The lack of herbicides with broad-spectrum activity and no crop injury was a consistent limitation in crop weed management. Multiple applications of numerous herbicides were routinely used to control a wide range of weed species detrimental to agronomic crops. Weed management tended to rely on preemergence—that is, herbicide applications were sprayed in response to expected weed infestations rather than in response to actual weeds present. Mechanical cultivation and hand weeding were often necessary to control weeds not controlled by herbicide applications.

The introduction of herbicide-tolerant crops has the potential of reducing the number of herbicide active ingredients used for weed management, reducing the number of herbicide applications made during a season, and increasing yield due to improved weed management and less crop injury. Transgenic crops that express tolerance to glyphosate, glufosinate and bromoxynil have been developed. These herbicides can now be sprayed on transgenic crops without inflicting damage on the crops while killing nearby weeds. From 1996 to 2001, herbicide tolerance was the most dominant trait introduced to commercially available transgenic crops, followed by insect resistance. In 2001, herbicide tolerance deployed in soybean, corn and cotton accounted for 77% of the 626,000 square kilometres planted to transgenic crops; Bt crops accounted for 15%; and "stacked genes" for herbicide tolerance and insect resistance used in both cotton and corn accounted for 8%.

### Production of Novel Substances in Crop Plants

Biotechnology is being applied for novel uses other than food. For example, oilseed can be modified to produce fatty acids for detergents, substitute fuels and petrochemicals. Potatoes, tomatoes, rice tobacco, lettuce, safflowers, and other plants have been genetically engineered to produce insulin and certain vaccines. If future clinical trials prove successful, the advantages of edible vaccines would be enormous, especially for developing countries.

The transgenic plants may be grown locally and cheaply. Homegrown vaccines would also avoid logistical and economic problems posed by having to transport traditional preparations over long distances and keeping them cold while in transit. And since they are edible, they will not need syringes, which are not only an additional expense in the traditional vaccine preparations but also a source of infections if contaminated.

In the case of insulin grown in transgenic plants, it is well-established that the gastrointestinal system breaks the protein down therefore this could not currently be administered as an edible protein. However, it might be produced at significantly lower cost than insulin produced in costly bioreactors. For example, Calgary, Canada-based SemBioSys Genetics, Inc. reports that its safflower-produced insulin will reduce unit costs by over 25% or more and approximates a reduction in the capital costs associated with building a commercial-scale insulin manufacturing facility of over $100 million, compared to traditional biomanufacturing facilities.

### *Criticism*

There is another side to the agricultural biotechnology issue. It includes increased herbicide usage and resultant herbicide resistance, "super weeds," residues on and in food crops, genetic contamination of non-GM crops which hurt organic and conventional farmers, etc.

## Biological Engineering

Biological engineering, biotechnological engineering or bioengineering (including biological systems engineering) is the application of concepts and methods of physics and mathematics to solve problems in life sciences, using engineering's own analytical and synthetical methodologies. In this context, while traditional engineering applies physical and mathematical sciences to analyse, design and manufacture inanimate tools, structures and processess, bioengineering uses the same sciences to study many aspects of living organisms. Usually it is used to analyse and solve problems related to human health.

Biological engineering is a science based discipline founded upon the biological sciences in the same way that chemical engineering, electrical engineering, and mechanical engineering are based upon chemistry, electricity and magnetism, and statics, respectively.

Biological Engineering can be differentiated from its roots of pure biology or classical engineering in the following way. Biological studies often follow a reductionist approach in viewing a system on its smallest possible scale which naturally leads toward tools such as functional genomics. Engineering approaches, using classical design perspectives, are constructionist, building new devices, approaches, and technologies from component concepts.

Biological engineering utilizes both of these methods in concert relying on reductionist approaches to define the fundamental units which are then commingled to generate something new. Although engineered biological systems have been used to manipulate information, construct materials, process chemicals, produce energy, provide food, and help maintain or enhance human health and our environment, our ability to quickly and reliably engineer biological systems that behave as expected remains less well developed than our mastery over mechanical and electrical systems.

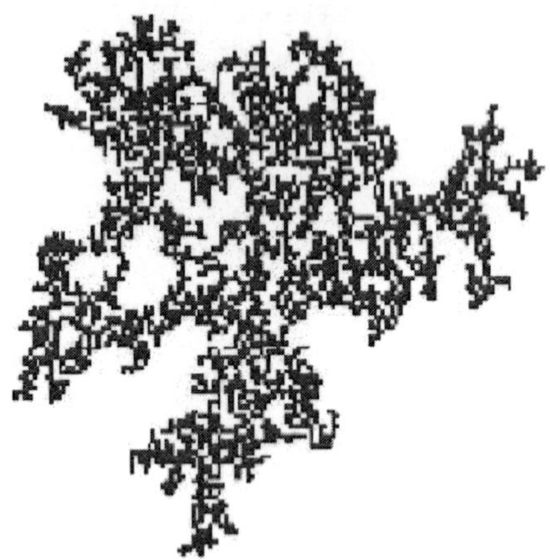

***Figure:*** *Modeling of the spread of disease using Cellular Automata and Nearest Neighbour Interactions*

The differentiation between Biological Engineering and overlap with Biomedical Engineering can be unclear, as many universities now use the terms "bioengineering" and "biomedical engineering" interchangeably. Some contend that Biological Engineering (like biotechnology) has a broader base which spans molecular methods (tends to emphasize the using of biological substances-applying engineering principles to molecular biology, biochemistry, microbiology, pharmacology, protein chemistry, cytology, immunology, neurobiology and neuroscience, cellular and tissue based methods (including devices and sensors), whole organisms (plants, animals), and up increasing length scales to ecosystems. Neither biological engineering nor biomedical engineering is wholly contained within the other, as there are non-biological products for medical needs and biological products for non-medical needs.

ABET, the U.S. based accreditation board for engineering B.S. programs, makes a distinction between Biomedical Engineering and Biological Engineering; however, the differences are quite small. Biomedical engineers must have life science courses that include human physiology and have experience in performing measurements on living systems while biological engineers must have life science courses (which may or may not include physiology) and experience in making measurements not specifically on living systems. Foundational engineering courses are often the same and include thermodynamics, fluid and mechanical dynamics, kinetics, electronics, and materials properties.

The word bioengineering was coined by British scientist and broadcaster Heinz Wolff in 1954. The term bioengineering is also used to describe the use of vegetation in civil engineering construction. The term bioengineering may also be applied to environmental modifications such as surface soil protection, slope stabilisation, watercourse and shoreline protection, windbreaks, vegetation barriers including noise barriers and visual screens, and the ecological enhancement of an area. The first biological engineering program was created at Mississippi State University in 1967, making it the first Biological Engineering curriculum in the United States. More recent programs have been launched at MIT and Utah State University.

*Biological Engineers* or *bioengineers* are engineers who use the principles of biology and the tools of engineering to create usable, tangible products. Biological Engineering employs knowledge and expertise from a number of pure and applied sciences, such as mass and heat transfer, kinetics, biocatalysts, biomechanics, bioinformatics, separation and purification processes, bioreactor design, surface science, fluid mechanics, thermodynamics, and polymer science. It is used in the design of medical devices, diagnostic equipment, biocompatible materials, renewable bioenergy, ecological engineering, and other areas that improve the living standards of societies.

In general, biological engineers attempt to either mimic biological systems in order to create products or modify and control biological systems so that they can replace, augment, or sustain chemical and mechanical processes. Bioengineers can apply their expertise to other applications of engineering and biotechnology, including genetic modification of plants and microorganisms, bioprocess engineering, and biocatalysis.

Because other engineering disciplines also address living organisms (e.g., prosthetics in mechanical engineering), the term biological engineering can be applied more broadly to include agricultural engineering and biotechnology. In fact, many old agricultural engineering departments in universities over the world have rebranded themselves as agricultural and biological engineering or agricultural and biosystems engineering. Biological engineering is also called bioengineering by some colleges and Biomedical engineering is called Bioengineering by others, and is a rapidly developing field with fluid categorization. The Main Fields of Bioengineering may be categorised as:

- Bioprocess Engineering: Bioprocess Design, Biocatalysis, Bioseparation, Bioinformatics, Bioenergy
- Genetic Engineering: Synthetic Biology, Horizontal gene transfer.
- Cellular Engineering: Cell Engineering, Tissue Culture Engineering, Metabolic Engineering.
- Biomedical Engineering: Biomedical technology, Biomedical Diagnostics, Biomedical Therapy, Biomechanics, Biomaterials.

## Microbial Biodegradation

Interest in the microbial biodegradation of pollutants has intensified in recent years as humanity strives to find sustainable ways to cleanup contaminated environments. These bioremediation and biotransformation methods endeavour to harness the astonishing, naturally occurring, ability of microbial xenobiotic metabolism to degrade, transform or accumulate a huge range of compounds including hydrocarbons (e.g. oil), polychlorinated biphenyls (PCBs), polyaromatic hydrocarbons (PAHs), heterocyclic compounds (such as pyridine or quinoline), pharmaceutical substances, radionuclides and metals. Major methodological breakthroughs in recent years have enabled detailed genomic, metagenomic, proteomic, bioinformatic and other high-throughput analyses of environmentally relevant microorganisms providing unprecedented insights into key biodegradative pathways and the ability of organisms to adapt to changing environmental conditions.

The elimination of a wide range of pollutants and wastes from the environment is an absolute requirement to promote a sustainable development of our society with low environmental impact. Biological processes play a major role in the removal of contaminants and they take advantage of the astonishing catabolic versatility of microorganisms to degrade/convert such compounds. New methodological breakthroughs in sequencing, genomics, proteomics, bioinformatics and imaging are producing vast amounts of information.

In the field of Environmental Microbiology, genome-based global studies open a new era providing unprecedented *in silico* views of metabolic and regulatory networks, as well as clues to the evolution of degradation pathways and to the molecular adaptation strategies to changing environmental conditions. Functional genomic and metagenomic approaches are increasing our understanding of the relative importance of different pathways and regulatory networks to carbon flux in particular environments and for particular compounds and they will certainly accelerate the development of bioremediation technologies and biotransformation processes.

### *Aerobic Biodegradation of Pollutants*

The burgeoning amount of bacterial genomic data provides unparalleled opportunities for understanding the genetic and molecular bases of the degradation of organic pollutants. Aromatic compounds are among the most recalcitrant of these pollutants and lessons can be learned from the recent genomic studies of *Burkholderia xenovorans* LB400 and *Rhodococcus* sp. strain RHA1, two of the largest bacterial genomes completely sequenced to date. These studies have helped expand our understanding of bacterial catabolism, non-catabolic physiological adaptation to organic compounds, and the evolution of large bacterial genomes. First, the metabolic pathways from phylogenetically diverse isolates are very similar with respect to overall organization.

Thus, as originally noted in pseudomonads, a large number of "peripheral aromatic" pathways funnel a range of natural and xenobiotic compounds into a restricted number of "central aromatic" pathways. Nevertheless, these pathways are genetically organized in genus-specific fashions, as exemplified by the b-ketoadipate and Paa pathways. Comparative genomic studies further reveal that some pathways are more widespread than initially thought.

Thus, the Box and Paa pathways illustrate the prevalence of non-oxygenolytic ring-cleavage strategies in aerobic aromatic degradation processes. Functional genomic studies have been useful in establishing that even organisms harboring high numbers of homologous enzymes seem to contain few examples of true redundancy. For example, the multiplicity of ring-cleaving dioxygenases in certain rhodococcal isolates may be attributed to the cryptic aromatic catabolism of different terpenoids and steroids.

Finally, analyses have indicated that recent genetic flux appears to have played a more significant role in the evolution of some large genomes, such as LB400's, than others. However, the emerging trend is that the large gene repertoires of potent pollutant degraders such as LB400 and RHA1 have evolved principally through more ancient processes. That this is true in such phylogenetically diverse species is remarkable and further suggests the ancient origin of this catabolic capacity.

### *Anaerobic Biodegradation of Pollutants*

Anaerobic microbial mineralization of recalcitrant organic pollutants is of great environmental significance and involves intriguing novel biochemical reactions. In particular, hydrocarbons and halogenated compounds have long been doubted to be degradable in the absence of oxygen, but the isolation of hitherto unknown anaerobic hydrocarbon-degrading and reductively dehalogenating bacteria during the last decades provided ultimate proof for these processes in nature.

Many novel biochemical reactions were discovered enabling the respective metabolic pathways, but progress in the molecular understanding of these bacteria was rather slow, since genetic systems are not readily applicable for most of them. However, with the increasing application of genomics in the field of environmental microbiology, a new and promising perspective is now at hand to obtain molecular insights into these new metabolic properties. Several complete genome sequences were determined during the last few years from bacteria capable of anaerobic organic pollutant degradation. The ~4.7 Mb genome of the facultative denitrifying *Aromatoleum aromaticum* strain EbN1 was the first to be determined for an anaerobic hydrocarbon degrader (using toluene or ethylbenzene as substrates). The genome sequence revealed about two dozen gene clusters (including several paralogs) coding for a complex catabolic network for anaerobic and aerobic degradation of aromatic compounds.

The genome sequence forms the basis for current detailed studies on regulation of pathways and enzyme structures. Further genomes of anaerobic hydrocarbon degrading bacteria were recently completed for the iron-reducing species *Geobacter metallireducens* (accession nr. NC_007517) and the perchlorate-reducing *Dechloromonas aromatica* (accession nr. NC_007298), but these are not yet evaluated in formal publications.

Complete genomes were also determined for bacteria capable of anaerobic degradation of halogenated hydrocarbons by halorespiration: the ~1.4 Mb genomes of *Dehalococcoides ethenogenes* strain 195 and *Dehalococcoides* sp. strain CBDB1 and the ~5.7 Mb genome of *Desulfitobacterium hafniense* strain Y51. Characteristic for all these bacteria is the presence of multiple paralogous genes for reductive dehalogenases, implicating a wider dehalogenating spectrum of the organisms than previously known. Moreover, genome sequences provided unprecedented insights into the evolution of reductive dehalogenation and differing strategies for niche adaptation.

Recently, it has become apparent that some organisms, including *Desulfitobacterium chlororespirans*, originally evaluated for halorespiration on chlorophenols, can also use certain brominated compounds, such as the herbicide bromoxynil and its major metabolite as electron acceptors for growth. Iodinated compounds may be dehalogenated as well, though the process may not satisfy the need for an electron acceptor.

### Bioavailability, Chemotaxis, and Transport of Pollutants

Bioavailability, or the amount of a substance that is physiochemically accessible to microorganisms is a key factor in the efficient biodegradation of pollutants. O'Loughlin et al (2000) showed that, with the exception of kaolinite clay, most soil clays and cation exchange resins attenuated biodegradation of 2-picoline by *Arthrobacter* sp. strain R1, as a result of adsorption of the substrate to the clays. Chemotaxis, or the directed movement of motile organisms towards or away from chemicals in the environment is an important physiological response that may contribute to effective catabolism of molecules in the environment. In addition, mechanisms for the intracellular accumulation of aromatic molecules via various transport mechanisms are also important.

### Oil Biodegradation

Petroleum oil contains aromatic compounds that are toxic for most life forms. Episodic and chronic pollution of the environment by oil causes major ecological perturbations. Marine environments are especially vulnerable since oil spills of coastal regions and the open sea are poorly containable and mitigation is difficult. In addition to pollution through human activities, about 250 million litres of petroleum enter the marine environment every year from natural seepages.

Despite its toxicity, a considerable fraction of petroleum oil entering marine systems is eliminated by the hydrocarbon-degrading activities of microbial communities, in particular by a remarkable recently discovered group of specialists, the so-called hydrocarbonoclastic bacteria (HCB). *Alcanivorax borkumensis* was the first HCB to have its genome sequenced. In addition to hydrocarbons, crude oil often contains various heterocyclic compounds, such as pyridine, which appear to be degraded by similar, though separate mechanisms than hydrocarbons.

### Analysis of Waste Biotreatment

Sustainable development requires the promotion of environmental management and a constant search for new technologies to treat vast quantities of wastes generated by increasing anthropogenic activities. Biotreatment, the processing of wastes using living organisms, is an environmentally friendly, relatively simple and cost-effective alternative to physico-chemical clean-up options.

Confined environments, such as bioreactors, have been engineered to overcome the physical, chemical and biological limiting factors of biotreatment processes in highly controlled systems. The great versatility in the design of confined environments allows the treatment of a wide range of wastes under optimized conditions.

To perform a correct assessment, it is necessary to consider various microorganisms having a variety of genomes and expressed transcripts and proteins. A great number of analyses are often required. Using traditional genomic techniques, such assessments are limited and time-consuming. However, several high-throughput techniques originally developed for medical studies can be applied to assess biotreatment in confined environments.

### Metabolic Engineering and Biocatalytic Applications

The study of the fate of persistent organic chemicals in the environment has revealed a large reservoir of enzymatic reactions with a large potential in preparative organic synthesis, which has already been exploited for a number of oxygenases on pilot and even on industrial scale. Novel catalysts can be obtained from metagenomic libraries and DNA sequence based approaches.

Our increasing capabilities in adapting the catalysts to specific reactions and process requirements by rational and random mutagenesis broadens the scope for application in the fine chemical industry, but also in the field of biodegradation. In many cases, these catalysts need to be exploited in whole cell bioconversions or in fermentations, calling for system-wide approaches to understanding strain physiology and metabolism and rational approaches to the engineering of whole cells as they are increasingly put forward in the area of systems biotechnology and synthetic biology.

### Fungal Biodegradation

In the ecosystem, different substrates are attacked at different rates by consortia of organisms from different kingdoms. *Aspergillus* and other moulds play an important role in these consortia because they are adept at recycling starches, hemicelluloses, celluloses, pectins and other sugar polymers.

Some aspergilli are capable of degrading more refractory compounds such as fats, oils, chitin, and keratin. Maximum decomposition occurs when there is sufficient nitrogen, phosphorus and other essential inorganic nutrients. Fungi also provide food for many soil organisms.

For *Aspergillus* the process of degradation is the means of obtaining nutrients. When these moulds degrade human-made substrates, the process usually is called biodeterioration. Both paper and textiles (cotton, jute, and linen) are particularly vulnerable to *Aspergillus* degradation. Our artistic heritage is also subject to *Aspergillus* assault. To give but one example, after Florence in Italy flooded in 1969, 74% of the isolates from a damaged Ghirlandaio fresco in the Ognissanti church were *Aspergillus versicolor*.

### Biotechnology Regulations

The National Institute of Health was the first federal agency to assume regulatory responsibility in the United States. The Recombinant DNA Advisory Committee of the NIH published guidelines for working with recombinant DNA and recombinant organisms in the laboratory.

Nowadays, the agencies that are responsible for the biotechnology regulation are: US Department of Agriculture (USDA) that regulates plant pests and medical preparation from living organisms, Environmental Protection Agency (EPA) that regulates pesticides and herbicides, and the Food and Drug Administration (FDA) which ensures that the food and drug products are safe and effective

### Education

In 1988, after prompting from the United States Congress, the National Institute of General Medical Sciences (National Institutes of Health) instituted a funding mechanism for biotechnology training. Universities nationwide compete for these funds to establish Biotechnology Training Programs (BTPs).

Each successful application is generally funded for five years then must be competitively renewed. Graduate students in turn compete for acceptance into a BTP; if accepted then stipend, tuition and health insurance support is provided for two or three years during the course of their PhD thesis work. Nineteen institutions offer NIGMS supported BTPs. Biotechnology training is also offered at the undergraduate level and in community colleges.

# Chapter 3

# Mathematical Modelling of Gene Regulation

The genome of an organism plays a crucial role in the regulation of cellular processes like adaptation, differentiation, development and maintenance. The main role in these processes is the information flow from the genome to the proteins, i.e., protein synthesis. Proteins synthesized by transcription of genes control almost every cellular function, but here we underline two important functions related to the regulation mechanisms.

According to the current metabolic state, a cell activates or deactivates some particular biochemical reactions. For example, in the absence of dietary glucose, glucose availability should be maintained by activating related biochemical reactions. These reactions are catalysed by enzymes most of which are proteins. Enzymes stabilize the transition states of the reactions by increasing the rate of the reactions. Biochemical reactions without enzymes usually take a very long time compared to the catalysed reactions.

Here, the most significant question is how cells adjust the expression of genes in response to different environmental conditions. The answer is gene regulation. Gene regulation is introduced by Jacob, Lwoff and Monod in, where they discovered that proteins bind to regulatory regions of other genes and function as transcription factors. This discovery constitutes the second important function related to genes. For example, phosphofructokinase is an enzyme which controls the flux of metabolites through the glycolytic pathway.

These roles imply that proteins generate a regulation network where they not only regulate the cellular processes, but the availability of each other as well. Although we define the regulation network in terms of proteins, we underline that genes constitute a basis for all regulatory factors in biological processes since proteins are encoded by genes. Thus, a regulation network should be constructed by means of gene interactions.

Nowadays, biochips are mentioned and enthusiastically praised in television, magazines and newspapers. There is a big public attention and fashion which almost expect medical wonders from this technology, regarded as a breakthrough combination of biology and computer engineering. However, what is mathematically behind such high expectations? The emergence of new high-throughput functional genomics technologies is a crucial driving factor in our ability to understand gene interactions Microarray technologies provide us with gene expression profiles which are reflections of the dynamical Behaviour of biological systems and the underlying regulatory network of interacting genes.

Achieving an understanding of gene regulation requires more than merely collecting the gene expression profiles. The dynamics of the underlying gene interaction network, chemical structure of the enzymes, allosteric regulations, enzymatic reaction rates, environmental factors that affect the regulation mechanisms should be studied to gain such an understanding. Mathematical modelling of such a complex system is a key issue in bioinformatics that deals with the time-series gene expression patterns to infer the underlying gene regulatory network.

## Current Mathematical Research Studies

The idea of mathematical description of biological organisms was first introduced by Turing in 1952. Turing stated that the change of state of a cell is the sum of all forces acting on that cell, which is a foundation for regulation mechanisms. Until 1952, different model classes reflecting the dynamics of genetic regulatory networks have been studied, including Boolean networks, Bayesian networks, stochastic models and differential equations.

The most widespread formalism is the latter modelling approach, i.e., differential equations, because of the qualitative dynamics of regulatory interactions. Regulation mechanisms are composed of interconnected regulatory pathways, and most of the factors which affect the dynamical Behaviour have not been determined yet.

This mystery distinguishes the modelling approach from the others which either make strong assumptions on the structure and dynamics of regulatory interactions or totally disregard the dynamical Behaviour of the genetic regulatory systems.

Another attractive feature of using differential equations is that they evolve in continuous time which is appropriate for the nature of the regulation networks. Response to a constant environmental data in a stable level of gene expressions and differential equations can determine the steady states analytically. This is another critical advance of using differential equations in modelling. Mathematically modelling such a system with differential equations combined with the theories of inverse problems and statistical learning constitutes our wish to model and anticipate the gene interactions.

Although we underline the advantages of using differential equations to model the dynamical Behaviour of gene expression patterns, two critical disadvantages of the model come into existence, the computation time and accuracy of the system. The interconnected structure and global view of the biological system is protected by using differential equations which makes the computational drawbacks less important. But some improvements can effectively reduce the computation time.

Since gene regulatory matrices are essentially sparse, the matrix can be partitioned into appropriate subsystems and this partitioning enables parallel processing. Accuracy deficiency of the system results from the insufficient experimental data. When sufficient source of data is available, differential systems can be solved numerically to any desired precision by adjusting the prediction error between real and approximated data.

### Brief Description of the Study

The metabolic state of a cell is usually regulated by gene and enzyme regulation mechanisms in the cell which ensures that the current metabolic state can be used to predict the next metabolic states.

Since only a limited portion of the gene interactions are known, we do not have sufficient information to build a first principle model. Our approach, also being preferred by most of the functional genomic modelling researchers, is selecting a model class and inferring the parameters from empirical data. Here, selecting the appropriate model class is the most crucial step. Firstly, estimation of parameters characterizing the dynamics of the system should have a unique solution. Secondly, the model should be able to approach the dynamics of the system as accurate as possible. For example, a linear differential system is not sufficient to seize the nonlinearities in the Behaviour of the system. On the other hand, the exact model will be included within the class of "any possible differential equations" but it cannot be distinguished from the infinitely many other solutions fitting to a particular data.

## Genetic Material

Organisms contain a genetic material which controls both phenotypic and genotypic Traits. There are four major characteristics of this genetic material stated: Replication, variation of mutation, storage and expression of information. The main characteristic related to genotype is replication which is a source of heredity. Genetic material is used as a template for the synthesis of a new strand, i.e., information encoded in genetic material is fully transferred to the offspring by replication.

Expression of information is a foundation for phenotypic processes. A phenotypic process requires some proteins to be synthesized. Information needed for each protein is encoded in some expressive part of the genetic material, called gene. Genes which are capable of information storage, serve as a template for the synthesis of related proteins, regulate transcription and make the phenotypic processes possible. In addition to the role in phenotypic processes, proteins have various functions in a cell cycle. However, in this study, we concentrate on the gene regulation mechanism in phenotypic processes.

## Regulation of Transcription

In order to respond environmental changes, growth and differentiation, organisms regulate the catalytic activities of enzymes by controlling the availability and activity of enzymes.

Enzyme availability is regulated by the rate of synthesis and degradation of an enzyme. Allosteric mechanisms, the structural alterations which influence the enzyme's binding affinity, alter the enzymatic activity. Donald et al. and Guyton et al. explain the gene and enzyme regulation mechanisms in detail. In the following sections, we give a brief summary of the regulation mechanisms.

The regulation mechanisms not only include gene and enzyme regulations, but external stimulus, e.g., hormones, as well. For example, insulin and glucagon are the two major hormones that regulate glycolysis pathway.

Glucagon acts to maintain glucose availability in the absence of dietary glucose where insulin is an anabolic hormone that acts in the opposite way. Although regulation networks consists of very complex and integrated regulation mechanisms, we underline that gene regulation is a very important factor in regulation mechanisms since gene transcription regulates transcription factors, which in turn regulate gene transcription.

### Enzyme Regulation

The first regulator mechanism in a cell is enzyme regulation. Main characteristics related to enzyme regulation are feedback loops in biochemical reactions and selfregulating substances. In biochemical synthesis, usually the synthesized substance can deactivate the first necessary enzyme for the reactions. This inhibition makes a decrease in concentrations of the intermediate substances by deactivating transcription and regulates the synthesis. In addition to this, some of the biochemical substances also have an ability to inhibit the enzymes which syntheses themselves.

### Gene Regulation

Expression of a gene depends on the activation of operon sequences. Activation and degradation of operon sequences are regulated by the promotor region. There are two main regions in promoter: repressor and activator operators. A repressor operator is critical in inhibition of protein synthesis because, if a repressor protein binds to this region, protein synthesis will be deactivated. There are activator proteins in the cell to break this bond and activate the transcription. An activator operator in promoter helps RNA polymerase to bind to promoter region, activates the operon and makes transcription possible.

If the synthesized material quantity reaches a critical value, the operon which is responsible for the synthesis of this material is inhibited by a negative feedback mechanism, either breaking the bonds between activator protein and operator, or binding repressor protein to repressor operator. In any case, operon is deactivated and protein synthesis is controlled.

Synthesis and regulation mechanisms in complex systems lead to hierarchical organization, differentiation and increased functionality. In prokaryotic genomes, genes with related functions are usually located in the same operon, as shown in the figure and transcribed together, while in eukaryotes most of the protein coding genes are transcribed individually. We will analyse how the complexity of organisms, i.e., the complexity of synthesis and regulation mechanisms, affects mathematical modelling approach in the coming sections.

If we have an information about the expression levels of genes, this information provides us an understanding about the gene interactions, i.e., gene regulation mechanisms. DNA arrays technology enables gene expression profiling, measuring the level of the mRNA gene products of a living cell simultaneously. The next section summarizes this technology and discusses the importance of the gene expression profiling.

## DNA Microarray Experiments

### *Purpose of the Experiments*

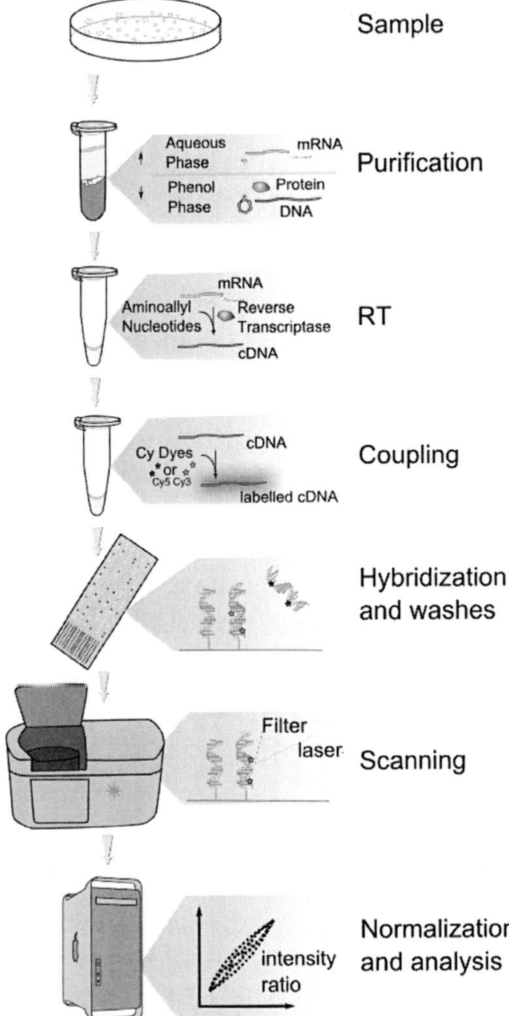


***Figure:*** *Steps involved in a microarray experiment (some steps omitted)*

DNA microarrays are designed to observe the gene expression of an experimental cell compared to a reference cell. For example, adding a suitable nutrient to an experimental cell makes a metabolic shift in the cell, by inhibiting or stimulating expressions levels of the genes, while metabolic state of the reference cell remains the same. After adding a nutrient, transcription rate of Gene A is lowered, while the expression level of Gene B becomes increased and Gene C does not give a response to the environmental change.

### *Design and Implementation*

There are several steps in design and implementation of a DNA microarray experiment. Array preparation is the first step in DNA microarray experiments. Suitable cDNA sequences which are obtained from mRNAs of experimental and reference cells, are selected and deposited on distinct glass arrays. Before printing, the glass arrays are treated to enhance binding of the cDNA to the glass surface by air drying and UV-crosslinking procedure, which increases fixation of the cDNA probes to the glass. The last step in array preparation is heating the arrays to separate strands of cDNA.

The second step is the critical step in the experiments. cDNAs from the experimental and reference samples are labelled with different fluorescent dyes, mixed, and hybridized to probes on the array. Hybridization refers to the binding of two cDNA strands by base pairing. After a sufficient time for hybridization and a series of washes to eliminate all unbounded target cDNAs and solution, each probe shows the expression level changes of a gene in experimental cell.

The last step of the experiments includes image processing techniques. The array is scanned for image processing analysis which is required to extract the numerical data. This process involves estimating the location of the spot on the array and, then, measuring the spot intensity as well as the background intensity based on the area outside the spot.

DNA microarray experiments provide us time-series experimental data for expression levels of genes, which show the relation between genotype and phenotype and, consequently, helps understanding biological processes. Systems ranging from gene regulation, to development, evolution and diseases, to differentiation as well as annotation of gene functions, identifying the effects of environment and life style, giving an insight into nutritional intake and therapeutic drugs, can be understood by DNA microarray experiments.

## Array Comparative Genomic Hybridization

Array comparative genomic hybridization (also CMA, Chromosomal Microarray Analysis, Microarray-based comparative genomic hybridization, array CGH, a-CGH, aCGH, or Virtual Karyotype) is a technique to detect genomic copy number variations at a higher resolution level than chromosome-based comparative genomic hybridization (CGH).

## Process

DNA from a test sample and normal reference sample are labelled differentially, using different fluorophores, and hybridized to several thousand probes. The probes are derived from most of the known genes and non-coding regions of the genome, printed on a glass slide. The ratio of the fluorescence intensity of the test to that of the reference DNA is then calculated, to measure the copy number changes for a particular location in the genome.

### *Efficiency*

Using this method, copy number changes at a level of 5-10 kilobases of DNA sequences can be detected. Today even high-resolution CGH (HR-CGH) arrays are accurate to detect structural variations (SV) at resolution of 200 bp. This method allows one to identify new recurrent chromosome changes such as:

- microdeletions
- Duplications.

In disease conditions such as cancer and birth defects due to chromosome aberrations.

### *Technical Considerations*

There are several requirements that are dependent on the application of aCGH:

- Complexity. Measurement becomes difficult in larger organisms because of decreasing partial concentrations of each portion of the sequence that is involved in the hybridization to the array element as the size of the genomes increase. This issue may be addressed by increasing the threshold in which one detects only larger increases in copy number of DNA extracted from cells, but this comes at the cost of increasing failure to detect low level gains and losses.
- Samples. Tissue specimens may contain heterogeneous cell populations, which may further decrease the ability to detect copy number change in genes in the aberrant tumour cells because the population may contain normal cells. Furthermore, the use of tissue from clinical specimens severely limit the amount of DNA available for analysis.

- Error tolerance. If the investigator is set to obtain a generalized description of aberrations that may occur in a set of samples, then errors in the detection may not be critical. However, the margin for error is drastically narrowed in a clinical setting, where an individual specimen is used to obtain specific information.

### *First Baby Born from the Procedure*

In September 2009, it was reported that the first baby born after screening with this procedure using polar body biopsy analysis, was Oliver, who was born in the UK to a 41-year-old woman. Of eight eggs that had been tested from the woman, only two were found to be chromosomally normal, and one of those two was used to make the embryo that became Oliver. Prior to Oliver being born, his mother had undergone 13 failed attempts at in vitro fertilization. The procedure was overseen by CARE Fertility.

## Comparative Genomic Hybridization

Comparative genomic hybridization (CGH) or Chromosomal Microarray Analysis (CMA) is a molecular-cytogenetic method for the analysis of copy number changes (gains/losses) in the DNA content of a given subject's DNA and often in tumour cells. CGH will detect only unbalanced chromosomal changes. Structural chromosome aberrations such as balanced reciprocal translocations or inversions cannot be detected, as they do not change the copy number. During the 1990s Thomas Cremer realized together with Peter Lichter the concept of comparative genomic hybridization to metaphase chromosomes and to a matrix with DNA spots representing specific genomic sites.

### *Method*

DNA from subject tissue and from normal control tissue are each labelled with different tags for later analysis by fluorescence. After mixing subject and reference DNA along with unlabeled human cot-1 DNA (placental DNA that is enriched for repetitive DNA sequences such as the Alu and Kpn family) to suppress repetitive DNA sequences, the mix is hybridized to normal metaphase chromosomes or, for array- or matrix-CGH, to a slide containing hundreds or thousands of defined DNA probes. Using epifluorescence microscopy and quantitative image analysis, regional differences in the fluorescence ratio of gains/losses vs. control DNA can be detected and used for identifying abnormal regions in the genome. Currently, strategically placed Oligos offer a resolution typically of 20-80 base pairs, as compared to the older BAC arrays offering resolution of 100kb. The (fluorescence) colour ratio along the chromosomes is used to evaluate regions of DNA gain or loss in the subject sample.

## Uses

### *Cancer*

The method is based on the hybridization of fluorescently labelled tumour DNA (frequently fluorescein (FITC)) and normal DNA (frequently rhodamine or Texas Red) to normal human metaphase preparations.

### *Other*

Aside from analysing cancer cells, the test is valuable at looking for previously unknown mutations that can lead to children with dysmorphic features, developmental delays, mental retardation, and autism. Preimplantation genetic diagnosis by CGH improves efficiency of in vitro fertilization by increasing embryo implantation rate and reducing multiple pregnancies and spontaneous miscarriages.

### *Limitations*

Chromosomal CGH is capable of detecting loss, gain and amplification of the copy number at the levels of chromosomes. However, it is considered that to detect a single copy loss the region must be at least 5-10 Mb in length. Detection of amplifications (e.g. tens or hundreds of copies of one or few neighbouring genes) is known to be sensitive down to less than 1 Mb. Therefore, one must take into consideration that while CGH is sensitive to specific types of copy number gains, the resolution of regional deletions is more limited.

The use of Array CGH overcomes many of these limitations, with improvement in resolution and dynamic range, in addition to direct mapping of aberrations to the genome sequence and improved throughput.

Due to the normalization to the most frequent ratio level as "normal", both CGH and array CGH do not provide information as to the ploidy. Since having a balanced DNA content, a tetraploid clone without further rearrangements would appear normal in CGH.

## Virtual Karyotype

Virtual Karyotype (also Array comparative genomic hybridization, CMA, Chromosomal Microarray Analysis, Microarray-based comparative genomic hybridization, array CGH, a-CGH, aCGH, or molecular karyotyping. If using SNP-based arrays, also SNP array karyotyping, molecular allelokaryotyping or SOMA) detects genomic copy number variations at a higher resolution level than conventional karyotyping or chromosome-based comparative genomic hybridization (CGH).

### Background

A karyotype is the characteristic chromosome complement of a eukaryote species. A karyotype is typically presented as an image of the chromosomes from a single cell arranged from largest (chromosome 1) to smallest (chromosome 22), with the sex chromosomes (X and/or Y) shown last. Historically, karyotypes have been obtained by staining cells after they have been chemically arrested during cell division. Karyotypes have been used for several decades to identify chromosomal abnormalities in both germline and cancer cells. Conventional karyotypes can assess the entire genome for changes in chromosome structure and number, but the resolution is relatively coarse with a detection limit of 5-10Mb.

### What is a Virtual Karyotype?

Recently, platforms for generating high-resolution karyotypes *in silico* from disrupted DNA have emerged, such as array comparative genomic hybridization (arrayCGH) and SNP arrays. Conceptually, the arrays are composed of hundreds to millions of probes which are complementary to a region of interest in the genome. The disrupted DNA from the test sample is fragmented, labelled, and hybridized to the array. The hybridization signal intensities for each probe are used by specialized software to generate a log2ratio of test/normal for each probe on the array. Knowing the address of each probe on the array and the address of each probe in the genome, the software lines up the probes in chromosomal order and reconstructs the genome *in silico*. Virtual karyotypes have dramatically higher resolution than conventional cytogenetics. The actual resolution will depend on the density of probes on the array. Currently, the Affymetrix SNP6.0 is the highest density commercially available array for virtual karyotyping applications. It contains 1.8 million polymorphic and non-polymorphic markers for a practical resolution of 10-20kb—about the size of a gene. This is approximately 1000-fold greater resolution than karyotypes obtained from conventional cytogenetics.

### How to get a Virtual Karyotype

Virtual karyotypes can be performed on germline samples for constitutional disorders, and clinical testing is available from dozens of CLIA certified laboratories (GeneTests.org). Virtual karyotyping can also be done on fresh or formalin-fixed paraffin-embedded tumours. CLIA-certified laboratories offering testing on tumours include Creighton Medical Laboratories (fresh and paraffin embedded tumour samples) and CombiMatrix Molecular Diagnostics (fresh tumour samples).

### Different Platforms for Virtual Karyotyping

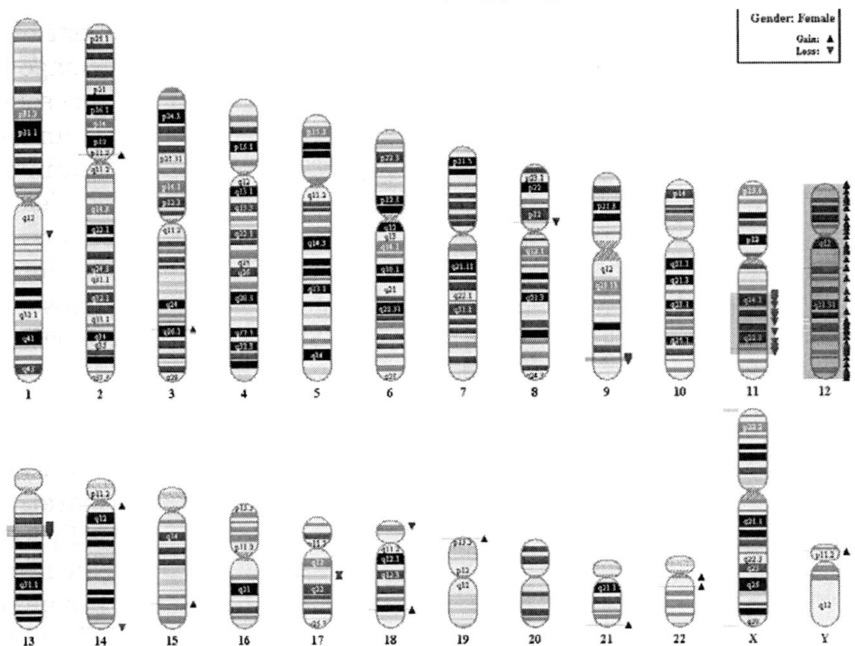


***Figure:*** *Virtual karyotype of a chronic lymphocytic leukemia sample using a SNP array.*

Array-based karyotyping can be done with several different platforms, both laboratory-developed and commercial. The arrays themselves can be genome-wide (probes distributed over the entire genome) or targeted (probes for genomic regions known to be involved in a specific disease) or a combination of both. Further, arrays used for karyotyping may use non-polymorphic probes, polymorphic probes (i.e., SNP-containing), or a combination of both. Non-polymorphic probes can provide only copy number information, while SNP arrays can provide both copy number and loss-of-heterozygosity (LOH) status in one assay. The probe types used for non-polymorphic arrays include cDNA, BAC clones (e.g., BlueGnome), and oligonucleotides (e.g., Agilent, Santa Clara, CA, USA or Nimblegen, Madison, WI, USA). Commercially available oligonucleotide SNP arrays can be solid phase (Affymetrix, Santa Clara, CA, USA) or bead-based (Illumina, SanDiego, CA, USA). Despite the diversity of platforms, ultimately they all use genomic DNA from disrupted cells to recreate a high resolution karyotype *in silico*.

The end product does not yet have a consistent name, and has been called virtual karyotyping, digital karyotyping, molecular allelokaryotyping, and molecular karyotyping. Other terms used to describe the arrays used for karyotyping include SOMA (SNP oligonucleotide microarrays) and CMA (chromosome microarray). Some consider all platforms to be a type of array comparative genomic hybridization (arrayCGH), while others reserve that term for two-dye methods, and still others segregate SNP arrays because they generate more and different information than two-dye arrayCGH methods.

### Applications

***Detecting Copy Number Changes:*** Copy number changes can be seen in both germline and tumour samples. Copy number changes can be detected by arrays with non-polymorphic probes, such as arrayCGH, and by SNP-based arrays. Human beings are diploid, so a normal copy number is always two for the non-sex chromosomes.

Deletions: A deletion is the loss of genetic material. The deletion can be heterozygous (copy number of 1) or homozygous (copy number of 0, nullisomy). Microdeletion syndromes are examples of constitutional disorders due to small deletions in germline DNA. Deletions in tumour cells may represent the inactivation of a tumour suppressor gene, and may have diagnostic, prognostic, or therapeutic implications.

Gains: A copy number gain represents the gain of genetic material. If the gain is of just one additional copy of a segment of DNA, it may be called a duplication. If there is one extra copy of an entire chromosome, it may be called a trisomy. Copy number gains in germline samples may be disease-associated or may be a benign copy number variant. When seen in tumour cells, they may have diagnostic, prognostic, or therapeutic implications.

***Loss of Heterozygosity (LOH), Autozygous Segments, and Uniparental Disomy:*** Autozygous segments and uniparental disomy (UPD) are diploid/'copy neutral' genetic findings and therefore are only detectable by SNP-based arrays. Both autozygous segments and UPD will show loss of heterozygosity (LOH) with a copy number of two by SNP array karyotyping. The term Runs of Homozgygosity (ROH), is a generic term that can be used for either autozygous segments or UPD.

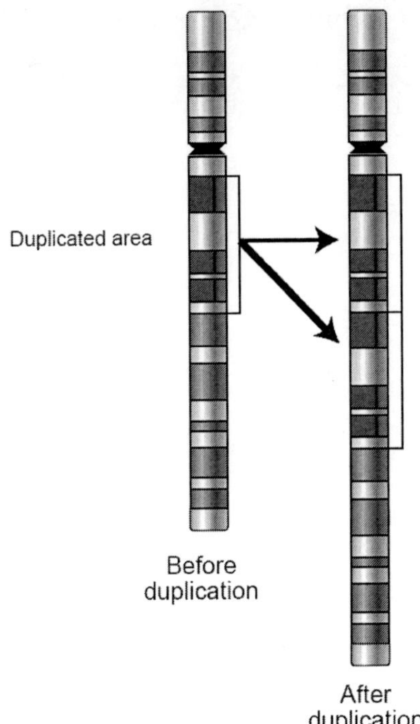


***Figure:** Schematic of a region of a chromosome before and after a duplication event*

***Autozygous segment:*** An autozygous segment is bi-parental and seen only in the germline. They are extended runs of homozygous markers in the genome, and they occur when an identical haplotype block is inherited from both parents. They are also called "identical by descent" (IBD) segments, and they can be used for homozygosity mapping. Uniparental Disomy: UPD occurs when both copies of a gene or genomic region are inherited from the same parent. This is uniparental, in contrast to autozygous segments which are bi-parental.

When present in the germline, they can be harmless or associated with disease, such as Prader-Willi or Angelman syndromes. Also in contrast to autozygosity, UPD can develop in tumour cells, and this is referred to as acquired UPD or copy neutral LOH in the literature.

Acquired UPD is quite common in both hematologic and solid tumours, and is reported to constitute 20 to 80% of the LOH seen in human tumours. Acquired UPD can serve as the 2nd hit in the Knudson Two Hit Hypothesis of Tumorigenesis, and thus can be the biological equivalent of a deletion. Because this type of lesion cannot be detected by arrayCGH, FISH, or conventional cytogenetics, SNP-based arrays are preferred for virtual karyotyping of tumours.

### Examples of Clinical Cancer Applications

A virtual karyotype can be generated from nearly any tumour, but the clinical meaning of the genomic aberrations identified are different for each tumour type.

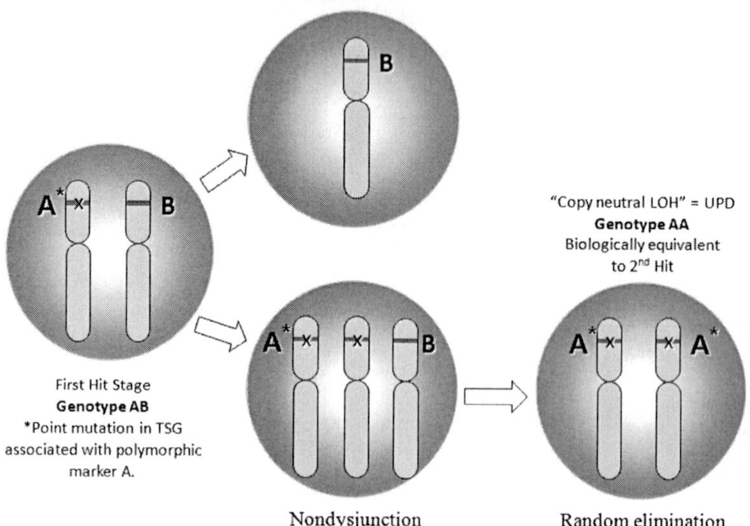


***Figure:*** *Copy neutral LOH/uniparental disomy*

Clinical utility varies and appropriateness is best determined by an oncologist or pathologist in consultation with the laboratory director of the lab performing the virtual karyotype. Below are examples of types of cancers where the clinical implications of specific genomic aberrations are well established. This list is representative, not exhaustive. The web site for the Cytogenetics Laboratory at Wisconsin State Laboratory of Hygiene has additional examples of clinically relevant genetic changes that are readily detectable by virtual karyotyping.

### Neuroblastoma

Based on a series of 493 neuroblastoma samples, it has been reported that overall genomic pattern, as tested by array-based karyotyping, is a predictor of outcome in neuroblastoma:

- Tumours presenting exclusively with whole chromosome copy number changes were associated with excellent survival.
- Tumours presenting with any kind of segmental chromosome copy number changes were associated with a high risk of relapse.

- Within tumours showing segmental alterations, additional independent predictors of decreased overall survival were MYCN amplification, 1p and 11q deletions, and 1q gain.

Earlier publications categorized neuroblastomas into three major subtypes based on cytogenetic profiles:

- Subtype 1: favourable neuroblastoma with near triploidy and a predominance of numerical gains and losses, mostly representing non-metastatic NB stages 1, 2 and 4S.
- Subtypes 2A and 2B: found in unfavourable widespread neuroblastoma, stages 3 and 4, with 11q loss and 17q gain without MYCN amplification (subtype 2A) or with MYCN amplification often together with 1p deletions and 17q gain (subtype 2B).

### *Wilms' Tumour*

Tumour-specific loss-of-heterozygosity (LOH) for chromosomes 1p and 16q identifies a subset of Wilms' tumour patients who have a significantly increased risk of relapse and death. LOH for these chromosomal regions can now be used as an independent prognostic factor together with disease stage to target intensity of treatment to risk of treatment failure.

### *Renal Cell Carcinoma*

Renal epithelial neoplasms have characteristic cytogenetic aberrations that can aid in classification.

- Clear cell carcinoma: loss of 3p
- Papillary carcinoma: trisomy 7 and 17
- Chromophobe carcinoma: hypodiploid with loss of chromosomes 1, 2, 6, 10, 13, 17, 21

Array-based karyotyping can be used to identify characteristic chromosomal aberrations in renal tumours with challenging morphology. Array-based karyotyping performs well on paraffin embedded tumours and is amenable to routine clinical use.

In addition, recent literature indicates that certain chromosomal aberrations are associated with outcome in specific subtypes of renal epithelial tumours. Clear cell renal carcinoma: del 9p and del 14q are poor prognostic indicators.

Papillary renal cell carcinoma: duplication of 1q marks fatal progression.

### *Chronic Lymphocytic Leukemia*

Array-based karyotyping is a cost-effective alternative to FISH for detecting chromosomal abnormalities in chronic lymphocytic leukemia (CLL). Several clinical validation studies have shown >95% concordance with the standard CLL FISH panel. In addition, many studies using array-based karyotyping have identified 'atypical deletions' missed by the standard FISH probes and acquired uniparental disomy at key loci for prognostic risk in CLL.,

Four main genetic aberrations are recognized in CLL cells that have a major impact on disease Behaviour.

1. Deletions of part of the short arm of chromosome 17 (del 17p) which target p53 are particularly deleterious. Patients with this abnormality have significantly short interval before they require therapy and a shorter survival. This abnormality is found in 5-10% of patients with CLL.
2. Deletions of the long arm on chromosome 11 (del 11q) are also unfavourable although not to the degree seen with del 17p. The abnormality targets the ATM gene and occurs infrequently in CLL (5-10%).
3. Trisomy 12, an additional chromosome 12, is a relatively frequent finding occurring in 20-25% of patients and imparts an intermediate prognosis.
4. Deletion of 13q14 (del 13q14) is the most common abnormality in CLL with roughly 50% of patients with cells containing this defect. When del 13q14 is seen in isolation, patients have the best prognosis and most will live many years, even decades, without the need for therapy.

### *Multiple Myeloma*

Avet-Loiseau, et al. in the Journal of Clinical Oncology, used SNP array karyotyping of 192 multiple myeloma (MM) samples to identify genetic lesions associated with prognosis, which were then validated in a separate cohort (n = 273). In MM, lack of a proliferative clone makes conventional cytogenetics informative in only ~30% of cases. FISH panels are useful in MM, but standard panels would not detect several key genetic abnormalities reported in this study.

1. Virtual karyotyping identified chromosomal abnormalities in 98% of MM cases
2. del(12p13.31)is an independent adverse marker
3. amp(5q31.1) is a favourable marker

4. The prognostic impact of amp(5q31.1) over-rides that of hyperdiploidy and also identifies patients who greatly benefit from high-dose therapy.

Array-based karytyping cannot detect balanced translocations, such as t(4;14) seen in ~15% of MM. Therefore, FISH for this translocation should also be performed if using SNP arrays to detect genome-wide copy number alterations of prognostic significance in MM.

### *Medulloblastoma*

Array-based karyotyping of 260 medulloblastomas by Pfister S, et al. resulted in the following clinical subgroups based on cytogenetic profiles:

- Poor prognosis: gain of 6q or amplification of MYC or MYCN
- Intermediate: gain of 17q or an i(17q) without gain of 6q or amplfication of MYC or MYCN
- Excellent prognosis: 6q and 17q balanced or 6q deletion.

### *Oligodendroglioma*

The 1p/19q co-deletion is considered a "genetic signature" of oligodendroglioma. Allelic losses on 1p and 19q, either separately or combined, are more common in classic oligodendrogliomas than in either astrocytomas or oligoastrocytomas. In one study, classic oligodendrogliomas showed 1p loss in 35 of 42 (83%) cases, 19q loss in 28 of 39 (72%), and these were combined in 27 of 39 (69%) cases; there was no significant difference in 1p/19q loss of heterozygosity status between low-grade and anaplastic oligodendrogliomas. 1p/19q co-deletion has been correlated with both chemosensitivity and improved prognosis in oligodendrogliomas. Most larger cancer treatment centres routinely check for the deletion of 1p/19q as part of the pathology report for oligodendrogliomas. The status of the 1p/19q loci can be detected by FISH or virtual karyotyping. Virtual karyotyping has the advantage of assessing the entire genome in one assay, as well as the 1p/19q loci. This allows assessment of other key loci in glial tumours, such as EGFR and TP53 copy number status.

Whereas the prognostic relevance of 1p and 19q deletions is well established for anaplastic oligodendrogliomas and mixed oligoastrocytomas, the prognostic relevance of the deletions for low-grade gliomas is more controversial. In terms of low-grade gliomas, a recent study also suggests that 1p/19q co-deletion may be associated with a (1;19)(q10;p10) translocation which, like the combined 1p/19q deletion, is associated with superior overall survival and progression-free survival in low-grade glioma patients. Oligodendrogliomas show only rarely mutations in the p53 gene, which is in contrast to other gliomas. Epidermal growth factor receptor amplification and whole 1p/19q codeletion are mutually exclusive and predictive of completely different outcomes, with EGFR amplification predicting poor prognosis.

### Glioblastoma

Yin et al studied 55 glioblastoma and 6 GBM cell lines using SNP array karyotyping. Acquired UPD was identified at 17p in 13/61 cases. A significantly shortened survival time was found in patients with 13q14 (RB) deletion or 17p13.1 (p53) deletion/acquired UPD. Taken together, these results suggest that this technique is a rapid, robust and inexpensive method to profile genome-wide abnormalities in GBM. Because SNP array karyotyping can be performed on paraffin embedded tumours, it is an attractive option when tumour cells fail to grow in culture for metaphase cytogenetics or when the desire for karyotyping arises after the specimen has been formalin fixed. The importance of detecting acquired UPD (copy neutral LOH) in glioblastoma:

- Of patients with 17p abnormality, ~50% were deletions and ~50% were aUPD
- Both 17p del and 17p UPD were associated with worse outcome.
- 9/13 had homozygous TP53 mutations underlying the 17p UPD.

In addition, in cases with uncertain grade by morphology, genomic profiling can assist in diagnosis.

- Concomitant gain of 7 and loss of 10 is essentially pathognomonic for GBM
- EGFR amplification, loss of PTEN (on 10q), and loss of p16 (on 9p) occur almost exclusively in glioblastoma and can provide means to distinguish anaplastic astrocytoma from glioblastoma.

### Acute Lymphoblastic Leukemia

Cytogenetics, the study of characteristic large changes in the chromosomes of cancer cells, has been increasingly recognized as an important predictor of outcome in acute lymphoblastic leukemia (ALL).

NB: Balanced translocations cannot be detected by array-based karyotyping.

Some cytogenetic subtypes have a worse prognosis than others. These include:

- A translocation between chromosomes 9 and 22, known as the Philadelphia chromosome, occurs in about 20% of adult and 5% in pediatric cases of ALL.

- A translocation between chromosomes 4 and 11 occurs in about 4% of cases and is most common in infants under 12 months.
- Not all translocations of chromosomes carry a poorer prognosis. Some translocations are relatively favourable. For example, Hyperdiploidy (>50 chromosomes) is a good prognostic factor.
- Genome-wide assessment of copy number changes can be done by conventional cytogenetics or virtual karyotyping. SNP array virtual karyotyping can detect copy number changes and LOH status, while arrayCGH can detect only copy number changes. Copy neutral LOH (acquired uniparental disomy) has been reported at key loci in ALL, such as CDKN2A gene at 9p, which have prognostic significance. SNP array virtual karyotyping can readily detect copy neutral LOH. Array CGH, FISH, and conventional cytogenetics cannot detect copy neutral LOH.

| ***Cytogenetic change*** | ***Risk category*** |
|---|---|
| Philadelphia chromosome | Poor prognosis |
| t(4;11)(q21;q23) | Poor prognosis |
| t(8;14)(q24.1;q32) | Poor prognosis |
| Complex karyotype (more than four abnormalities) | Poor prognosis |
| Low hypodiploidy or near triploidy | Poor prognosis |
| High hyperdiploidy | Good prognosis |
| del(9p) | Good prognosis |

Correlation of prognosis with bone marrow cytogenetic finding in acute lymphoblastic leukemia;

| ***Prognosis*** | ***Cytogenetic findings*** |
|---|---|
| Favourable | Hyperdiploidy > 50 ; t (12;21) |
| Intermediate | Hyperdioloidy 47 -50; Normal(diploidy); del (6q); Rearrangements of 8q24 |
| Unfavourable | Hypodiploidy-near haploidy; Near tetraploidy; del (17p); t (9;22); t (11q23) |

Unclassified ALL is considered to have an intermediate prognosis.

## *Myelodysplastic Syndrome*

Myelodysplastic syndrome (MDS) has remarkable clinical, morphological, and genetic heterogeneity. Cytogenetics play a decisive role in the World Health Organization's classification-based International Prognostic Scoring System (IPSS) for MDS.

- Good Prognosis: normal karyotype, isolated del(5q), isolated del(20q), -Y
- Poor Prognosis: complex abnormalities (ie, >=3 abnormalities), -7 or del(7q)
- Intermediate Prognosis: all other abnormalities, including trisomy 8 and del(11q).

In a comparison of metaphase cytogenetics, FISH panel, and SNP array karyotyping for MDS, it was found that each technique provided a similar diagnostic yield. No single method detected all defects, and detection rates improved by ~5% when all three methods were used.

Acquired UPD, which is not detectable by FISH or cytogenetics, has been reported at several key loci in MDS using SNP array karyotyping, including deletion of 7/7q.

### *Myeloproliferative Neoplasms/Myeloproliferative Disorders*

Philadelphia chromosome–negative myeloproliferative neoplasms (MPNs) including polycythemia vera,essential thrombocythemia, and primary myelofibrosis show an inherent tendency for transformation into leukemia (MPN-blast phase), which is accompanied by acquisition of additional genomic lesions. In a study of 159 cases, SNP-array analysis was able to capture practically all cytogenetic abnormalities and to uncover additional lesions with potentially important clinical implications.

- The number of genomic alterations was more than 2 to 3 times greater in the blast phase as in the chronic phase of the disease.
- Deletion of 17p (TP53)was significantly associated with prior exposure to hydroxyurea as well as a complex karyotype in samples with MPN-blast crisis. Interestingly, not only deletion, but also 17p copy neutral LOH, was associated with a complex karyotype, a poor prognostic marker in myeloid malignancies. Copy neutral LOH (acquired UPD)is readily detectably by SNP array karyotype, but not by cytogenetics, FISH, or arrayCGH.
- Blast phase patients with loss of chromosomal material on 7q showed poor survival. Loss of 7q is known to be predictive for rapid progression and poor response in AML therapy. MPN-blast phase patients with cytogenetically undetectable 7q copy neutral-LOH had comparable survival rates to those with 7/7q in their leukemic cells.
- 9p copy neutral-LOH with homozygous JAK2 mutation was also linked to an inferior outcome in MPN-blast crisis in comparison

with patients with either heterozygous JAK2V617F or wild-type JAK2. In contrast to LOH on 17p, the prognostic impact of 9pCNN-LOH was independent of established risk factors such as 7/7q, 5q, or complex karyotype.

### Colorectal Cancer

Identification of biomarkers in colorectal cancer is particularly important for patients with stage II disease, where less than 20% have tumour recurrence. 18q LOH is an established biomarker associated with high risk of tumour recurrence in stage II colon cancer.

Colorectal cancers are classified into specific tumour phenotypes based on molecular profiles which can be integrated with the results of other ancillary tests, such as microsatellite instability testing, IHC, and KRAS mutation status:

- Chromosomal instability (CIN) which have allelic imbalance at a number of chromosomal loci, including 5q, 8p, 17p, and 18q.
- Microsatellite instability (MSI) which tend to have diploid karyotypes.

### Malignant rhabdoid Tumours

Malignant rhabdoid tumours are rare, highly aggressive neoplasms found most commonly in infants and young children. Due to their heterogenous histologic features, diagnosis can often be difficult and misclassifications can occur. In these tumours, the INI1 gene (SMARCB1) on chromosome 22q functions as a classic tumour suppressor gene. Inactivation of INI1 can occur via deletion, mutation, or acquired UPD.

In a recent study, SNP array karyotyping identified deletions or LOH of 22q in 49/51 rhabdoid tumours. Of these, 14 were copy neutral LOH (or acquired UPD), which is detectable by SNP array karyotyping, but not by FISH, cytogenetics, or arrayCGH. MLPA detected a single exon homozygous deletion in one sample that was below the resolution of the SNP array.

SNP array karyotyping can be used to distinguish, for example, a medulloblastoma with an isochromosome 17q from a primary rhabdoid tumour with loss of 22q11.2. When indicated, molecular analysis of INI1 using MLPA and direct sequencing may then be employed. Once the tumour-associated changes are found, an analysis of germline DNA from the patient and the parents can be done to rule out an inherited or de novo germline mutation or deletion of INI1, so that appropriate recurrence risk assessments can be made.

### Uveal Melanoma

The most important genetic alteration associated with poor prognosis in uveal melanoma is loss of an entire copy of Chromosome 3 (Monosomy 3), which is strongly correlated with metastatic spread. Gains on chromosomes 6 and 8 are often used to refine the predictive value of the Monosomy 3 screen, with gain of 6p indicating a better prognosis and gain of 8q indicating a worse prognosis in disomy 3 tumours. In rare instances, monosomy 3 tumours may duplicate the remaining copy of the chromosome to return to a disomic state referred to as isodisomy. Isodisomy 3 is prognostically equivalent to monosomy 3, and both can be detected by tests for chromosome 3 loss of heterozygosity.

### Limitations

Unlike karyotypes obtained from conventional cytogenetics, virtual karyotypes are reconstructed by computer programs using signals obtained from *disrupted* DNA. In essence, the computer program will correct translocations when it lines up the signals in chromosomal order. Therefore, virtual karyotypes cannot detect balanced translocations and inversions. They also can only detect genetic aberrations in regions of the genome that are represented by probes on the array. In addition, virtual karyotypes generate a *relative* copy number normalized against a diploid genome, so tetraploid genomes will be condensed into a diploid space unless renormalization is performed. Renormalization requires an ancillary cell-based assay, such as FISH, if one is using arrayCGH. For karyotypes obtained from SNP-based arrays, tetraploidy can often be inferred from the maintenance of heterozygosity within a region of apparent copy number loss.

Low-level mosaicism or small subclones may not be detected by virtual karyotypes because the presence of normal cells in the sample will dampen the signal from the abnormal clone. The exact point of failure, in terms of the minimal percentage of neoplastic cells, will depend on the particular platform and algorithms used. Many copy number analysis software programs used to generate array-based karyotypes will falter with less than 25-30% tumour/abnormal cells in the sample. However, in oncology applications this limitation can be minimized by tumour enrichment strategies and/or software optimized for use with oncology samples. The analysis algorithms are evolving rapidly, and some are even designed to thrive on 'normal clone contamination', so it is anticipated that this limitation will continue to dissipate.

## Pharmacogenomics

Pharmacogenomics is the branch of pharmacology which deals with the influence of genetic variation on drug response in patients by correlating gene expression or single-nucleotide polymorphisms with a drug's efficacy or toxicity. By doing so, pharmacogenomics aims to develop rational means to optimise drug therapy, with respect to the patients' genotype, to ensure maximum efficacy with minimal adverse effects. Such approaches promise the advent of "personalized medicine"; in which drugs and drug combinations are optimized for each individual's unique genetic makeup.

Pharmacogenomics is the whole genome application of pharmacogenetics, which examines the single gene interactions with drugs. Pharmacogenomics is being used for all critical illnesses like cancer, cardio vascular disorders, HIV, tuberculosis, asthma, and diabetes. In cancer treatment, pharmacogenomics tests are used to identify which patient will have toxicity from commonly used cancer drugs and identify which patient will not respond to commonly used cancer drug. Over the last couple of years, pharmacogenomics is also known as companion diagnostics, meaning tests being bundled with drugs. Two good examples are K-ras test with cituximab and EGFR test with Gefitinib.

In cardio vascular disorders, the main concern is response to drugs including warfarin, clopidogrel, beta blockers, and statins.

A single-nucleotide polymorphism (SNP, pronounced *snip*) is a DNA sequence variation occurring when a single nucleotide — A, T, C, or G — in the genome (or other shared sequence) differs between members of a species or paired chromosomes in an individual. For example, two sequenced DNA fragments from different individuals, AAGCCTA to AAGCTTA, contain a difference in a single nucleotide. In this case we say that there are two *alleles*: C and T. Almost all common SNPs have only two alleles.

Within a population, SNPs can be assigned a minor allele frequency— the lowest allele frequency at a locus that is observed in a particular population. This is simply the lesser of the two allele frequencies for single-nucleotide polymorphisms. There are variations between human populations, so a SNP allele that is common in one geographical or ethnic group may be much rarer in another.

## Types of SNPs

Single nucleotide polymorphisms may fall within coding sequences of genes, non-coding regions of genes, or in the intergenic regions between genes. SNPs within a coding sequence will not necessarily change the amino acid sequence of the protein that is produced, due to degeneracy of the genetic code.

A SNP in which both forms lead to the same polypeptide sequence is termed *synonymous* (sometimes called a silent mutation) — if a different polypeptide sequence is produced they are *nonsynonymous*. A nonsynonymous change may either be missense or nonsense, where a missense change results in a different amino acid, while a nonsense change results in a premature stop codon. SNPs that are not in protein-coding regions may still have consequences for gene splicing, transcription factor binding, or the sequence of non-coding RNA.

### Use and Importance of SNPs

Variations in the DNA sequences of humans can affect how humans develop diseases and respond to pathogens, chemicals, drugs, vaccines, and other agents. SNPs are also thought to be key enablers in realizing the concept of personalized medicine. However, their greatest importance in biomedical research is for comparing regions of the genome between cohorts (such as with matched cohorts with and without a disease). The study of single-nucleotide polymorphisms is also important in crop and livestock breeding programs.

They are usually biallelic and thus easily assayed. SNPs do not necessarily function individually, rather, they work in coordination with other SNPs to manifest a disease condition as has been seen in osteoporosis.

### Examples

- rs6311 and rs6313 are SNPs in the HTR2A gene on human chromosome 13.
- A SNP in the *F5* gene causes a hypercoagulability disorder with the variant Factor V Leiden.
- rs3091244 is an example of a triallelic SNP in the CRP gene on human chromosome 1.
- TAS2R38 codes for PTC tasting ability, and contains 6 annotated SNPs.

### Databases

As there are for genes, there are also bioinformatics databases for SNPs. *dbSNP* is a SNP database from National Centre for Biotechnology Information (NCBI). *SNPedia* is a wiki-style database from a hybrid organization. The *OMIM* database describes the association between polymorphisms and, e.g., diseases in text form, *HGMD®* the Human Gene Mutation Database provides gene mutations causing or associated with human inherited diseases and functional SNPs, while HGVbaseG2P allows users to visually interrogate the actual summary-level association data.

## Nomenclature

The nomenclature for SNPs can be confusing: several variations can exist for an individual SNP and consensus has not yet been achieved. One approach is to write SNPs with a prefix, period and greater than sign showing the wild-type and altered nucleotide or amino acid; for example, c.76A>T.

What are the anticipated benefits of pharmacogenomics?

***More Powerful Medicines :*** Pharmaceutical companies will be able to create drugs based on the proteins, enzymes, and RNA molecules associated with genes and diseases. This will facilitate drug discovery and allow drug makers to produce a therapy more targeted to specific diseases. This accuracy not only will maximize therapeutic effects but also decrease damage to nearby healthy cells.

***Better, Safer Drugs the First Time :*** Instead of the standard trial-and-error method of matching patients with the right drugs, doctors will be able to analyse a patient's genetic profile and prescribe the best available drug therapy from the beginning. Not only will this take the guesswork out of finding the right drug, it will speed recovery time and increase safety as the likelihood of adverse reactions is eliminated. Pharmacogenomics has the potential to dramatically reduce the estimated 100,000 deaths and 2 million hospitalizations that occur each year in the United States as the result of adverse drug response.

***More Accurate Methods of Determining Appropriate Drug Dosages :*** Current methods of basing dosages on weight and age will be replaced with dosages based on a person's genetics —how well the body processes the medicine and the time it takes to metabolize it. This will maximize the therapy's value and decrease the likelihood of overdose.

***Advanced Screening for Disease :*** Knowing one's genetic code will allow a person to make adequate lifestyle and environmental changes at an early age so as to avoid or lessen the severity of a genetic disease. Likewise, advance knowledge of a particular disease susceptibility will allow careful monitoring, and treatments can be introduced at the most appropriate stage to maximize their therapy.

***Better Vaccines :*** Vaccines made of genetic material, either DNA or RNA, promise all the benefits of existing vaccines without all the risks. They will activate the immune system but will be unable to cause infections. They will be inexpensive, stable, easy to store, and capable of being engineered to carry several strains of a pathogen at once.

***Improvements in the Drug Discovery and Approval Process :*** Pharmaceutical companies will be able to discover potential therapies more easily using genome targets. Previously failed drug candidates may be revived as they are matched with the niche population they serve. The drug approval process should be facilitated as trials are targeted for specific genetic population groups —providing greater degrees of success. The cost and risk of clinical trials will be reduced by targeting only those persons capable of responding to a drug.

***Decrease in the Overall Cost of Health Care :*** Decreases in the number of adverse drug reactions, the number of failed drug trials, the time it takes to get a drug approved, the length of time patients are on medication, the number of medications patients must take to find an effective therapy, the effects of a disease on the body (through early detection), and an increase in the range of possible drug targets will promote a net decrease in the cost of health care.

### *Is Pharmacogenomics in use Today?*

To a limited degree. The cytochrome P450 (CYP) family of liver enzymes is responsible for breaking down more than 30 different classes of drugs. DNA variations in genes that code for these enzymes can influence their ability to metabolize certain drugs. Less active or inactive forms of CYP enzymes that are unable to break down and efficiently eliminate drugs from the body can cause drug overdose in patients. Today, clinical trials researchers use genetic tests for variations in cytochrome P450 genes to screen and monitor patients. In addition, many pharmaceutical companies screen their chemical compounds to see how well they are broken down by variant forms of CYP enzymes.

Another enzyme called TPMT (thiopurine methyltransferase) plays an important role in the chemotherapy treatment of a common childhood leukemia by breaking down a class of therapeutic compounds called thiopurines. A small percentage of Caucasians have genetic variants that prevent them from producing an active form of this protein. As a result, thiopurines elevate to toxic levels in the patient because the inactive form of TMPT is unable to break down the drug. Today, doctors can use a genetic test to screen patients for this deficiency, and the TMPT activity is monitored to determine appropriate thiopurine dosage levels.

### *What are some of the Barriers to Pharmacogenomics Progress?*

Pharmacogenomics is a developing research field that is still in its infancy. Several of the following barriers will have to be overcome before many pharmacogenomics benefits can be realized.

- Complexity of finding gene variations that affect drug response-Single nucleotide polymorphisms (SNPs) are DNA sequence variations that occur when a single nucleotide (A,T,C, or G) in the genome sequence is altered. SNPs occur every 100 to 300 bases along the 3-billion-base human genome, therefore millions of SNPs must be identified and analysed to determine their involvement (if any) in drug response. Further complicating the process is our limited knowledge of which genes are involved with each drug response. Since many genes are likely to influence responses, obtaining the big picture on the impact of gene variations is highly time-consuming and complicated.
- Limited drug alternatives-Only one or two approved drugs may be available for treatment of a particular condition. If patients have gene variations that prevent them using these drugs, they may be left without any alternatives for treatment.
- Disincentives for drug companies to make multiple pharmacogenomic products-Most pharmaceutical companies have been successful with their "one size fits all" approach to drug development. Since it costs hundreds of millions of dollars to bring a drug to market, will these companies be willing to develop alternative drugs that serve only a small portion of the population?
- Educating healthcare providers-Introducing multiple pharmacogenomic products to treat the same condition for different population subsets undoubtedly will complicate the process of prescribing and dispensing drugs. Physicians must execute an extra diagnostic step to determine which drug is best suited to each patient. To interpret the diagnostic accurately and recommend the best course of treatment for each patient, all prescribing physicians, regardless of specialty, will need a better understanding of genetics.

## Ethical Issues in Pharmacogenetics

Pharmacogenetics is the study of how genes influence an individual's response to drugs. Though the field would seem to be brand new, it is really half a century old. In the 1950's, scientists first identified deficiencies in enzymes that explained adverse reactions to drugs and that they could be inherited.

For example, early research showed that 10% of African American men serving in the Korean war became anemic after ingesting an anti-malarial drug, which rarely, if ever caused problems for Caucasian soldiers. To pinpoint the cause, it took years of study:

Trial and error used to determine drug development;

- The anemic reaction was determined to be caused by a variation of the G6PD gene, and the variation was found to be common among people of African descent but not so among Caucasians.
- It was later discovered that the normal form of the gene makes an enzyme that helps protect red blood cells against certain chemicals. Lacking that protective effect, those with the variant form are vulnerable to deleterious effects.
- Since that time, numerous other enzyme variants have been identified and found to cause adverse reactions. Such adverse effects were identified, until recently, by trial and error methods. Specifically, drugs were administered, and an individual's metabolism of that drug was tracked by recording the amount of by-product in their urine.

Genomic mapping eliminates trial and error drug trial methods.

The Human Genome Project has enabled us to identify the molecular composition of the enzymes in question so that we can study correlations between genotypic (gene trait) and phenotypic (physical trait) variability.

These advances will increasingly enable us to detect individuals who are likely to experience adverse reactions to medicines without having to use potentially dangerous methods of trial and error.

In coming years, we are likely to learn that particular single nucleotide polymorphisms (SNPs) are associated with sensitivities or resistances to chemical compounds in the environment. Scientists are now rushing to not only identify common SNPs, but to determine what drug effects can be correlated to them.

Pharmacogenomics is a recent offshoot of pharmacogenetics. Its scope is broader; for example, it attempts to understand not only the molecular composition of genetic variants associated with drug response but also the Behaviour of those variants, including how those genes affect drug receptor sites.

*Ethical Issue #1: "Good" or "Bad" Allocation of Scarce Resources?*

Critics say that genomic mapping is a waste of money and time.

Many believe that pharmacogenomics, like other new fields spawned by the Human Genome Project, represent a misallocation of resources. Rather than embark on learning how genes indicate a predisposition to disease and developing cures and enhancements, or experimenting with ways to change the human germ cell, global efforts should be spent on solving more urgent problems facing humanity, such as global famine or accessibility to potable water.

Others contend that pharmacogenomics, in particular, offers enormous potential for clinical benefits to patients as well as economic benefits for health care delivery. The arguments in favour include:

Every year, adverse reactions to drugs possibly kill 100,000 American patients.

- In the U.S. alone, adverse drug reactions are thought to KILL about 100,000 hospitalized patients annually. It is believed that many of these reactions are due to genetic variants and thus many of these deaths can be avoided by testing people for adverse drug response before giving them drugs. The science and technology for such tests, however, are in their infancy.

Over 2 million people have serious reactions to medication.

- Another 2.2 million incur serious, but non-fatal, reactions. Physicians, in view of their Hippocratic oath, are obligated to do no harm. Can this obligation be fulfilled when the information available to physicians about how particular medicines will fare in their patients is so meager? At present, physicians, generally have no way of knowing in advance whether the drug they prescribe will or will not cause an adverse effect in their patients.

'One-size- fits-all' medication can be dangerous.

- This situation is further compounded by the fact that most adverse drug reactions result from the fact that medicines are "a one-size-fits-all." In other words, although medicines are taken in different dosages depending on symptoms, patient age, weight and other clinical factors, these criteria may not be adequate to ensure that a particular medicine will be safe and effective for a particular individual. Until recently, there has been no alternative to either developing or prescribing medicines. Pharmacogenomics promises to take the guesswork out of developing and prescribing safe and effective drugs.

*Ethical Issue #2: What is a fair distribution of burdens and benefits in developing the field of pharmacogenomics?*

Monies and people (as research subjects and as researchers) will develop the field to the point that customized medicine will be possible. Who will benefit?

Designer drugs may be too costly at first for all to benefit.

- The availability of this new technology may be costly initially, and thus accessible only to those wealthy enough to pay for both the test and the designer drug best suited to them. Yet, the cost will likely diminish so as to become affordable to most. However, will lower costs influence a person to submit to the required genetic testing, thus creating threats, if not violations, to one's autonomy (the basic tenet of bioethics)?

Commercialization of research results may lead to conflict of interest.

- Researchers who have investments in companies competing in their field may be in a conflict of interest if they are conducting research for such a company. Substantial concerns about conflicts of interest as both a threat to quality research as well as to the well being of research subjects have abounded for decades.
- A recent study found that policies governing conflicts of interests at major medical institutions varied considerably in both disclosure requirements and the nature of permitted academic-industry relationship, thereby opening a door to the possibility that an interest in financial gain could overpower an interest in either achieving valid research or protecting the well-being of subjects.

Participants do not always benefit from successful outcomes of drug trials.

- Further, there are several examples in the history of medical research where the patient population standing to benefit from advances (i.e., people who have donated their time, bodies, and hearts to research, though compensated per standard National Institute of Health [NIH] terms), did not receive the anticipated medical benefits because new therapies were unaffordable, when they became commercially available, or not covered by insurers, as these two examples illustrate:
- Numerous sufferers of Gauchier Disease, who helped companies developed safe and effective treatment (clinical research), were denied access to treatments by insurance companies by refusing

to cover the high cost therapies. The patients couldn't afford to pay costs out of their own pocket.

- A Canavan's Disease Support Group has been instrumental in helping a company develop treatment by raising research funds as well as supplying researchers with willing research participants. The group is suing research facilities not for financial return on investment but for the opportunity to play an active role in furthering research/treatment goals.

*Ethical Issue #3: Will individualized medicine be used ethically?*

Knowing if a person will respond to a drug in ways that are safe and effective for that individual will enable patients to avoid medications that are dangerous or ineffective for them.

Gene profiling is not the only factor in creating designer drugs.

This is not to say that genes are the only key to cures. Environment plays a role, too. Dietary and lifestyle Behaviours are likely to still affect the safety and efficacy of medicines for particular individuals. As well, variation in drug response is not limited to micro polymorphisms. Environmental factors also play a role (such as sun exposure, drug/drug interaction, drug/food interaction). However, scientists are poised to uncover why the metabolism of particular individuals absorbs and dispels pharmaceuticals in a particular manner.

Consider the following *hypothetical* clinical scenario as illustrating some of the ethical issues that can arise in clinic:

Not all physicians will take advantage of new testing methods.

- A 42 year old man of Scandinavian descent presents to his physician with a general feeling of malaise.
- Five years previously he was diagnosed with high serum cholesterol, which he attempted to control with a regimen of exercise and dietary regulation, with no success. His physician then prescribed for him a drug therapy.
- Before agreeing to take the prescribed medication, the patient retrieved volumes of information from the Web, including but not limited to peer-reviewed journal articles about his condition and his physician's first choice drug.
- After six months of therapy there was only a modest lowering of cholesterol levels, so the medication was changed. After nine months on the second medication, there was still no marked effect.

- By the time the patient was able to see his physician again, a newer therapy had become available. This new drug had become the physician's favourite. The physician advised the patient to switch to this new drug, and the patient was eager to try it. Three weeks later, the patient came to see the physician to complain of continued malaise.

The patient may have been better served if he had undergone the following genetic tests, the results of which could have provided valuable management information:

Genetic testing can provide an array of diagnostic results.

- Test 1: a pre-dispositional test to determine whether the patient has a polymorphism associated with plaque development leading to coronary heart disease.
- Test 2: a test to see whether the patient has a polymorphism associated with a non-response to the medication (the newest medicine). A positive test 2 indicates that the patient lacks an enzyme needed to metabolize the drug. The absence of the enzyme means that the drug is dispelled from the body without absorption.
- Test 3: a test to see whether the patient has a polymorphism which indicates the presence of an enzyme responsible for metabolizing the dosage too slowly, making the drug in that dosage toxic to the patient.
- The rationale sequence of testing is 1-3.

If the patient tests negative, meaning that he does not have the polymorphism associated with plaque development, then his high cholesterol poses no health risk and medication to lower cholesterol levels are not indicated. If the patient tests positive, meaning that he does have the polymorphism, then he is predisposed to coronary heart disease (CAD) by virtue of being a plaque maker. In this case, cholesterol-lowering medication is indicated.

*Ethical Issue #4: Whose right predominates?*

The father of a research subject opened a letter addressed to his child and learned that his child had enrolled in a genetic research study.

*Whose privacy was invaded — the father's or the child's?*

- The letter indicated that for the purpose of research the research facility had obtained some of the father's medical records. The father objected to what apparently was non-consensual disclosure of his medical information, even for the purpose of obtaining an informative family history to be used to provide optimal care for the son/daughter.

- Outraged, the father phoned the Office for Human Research Protections (OHRP), of the U.S. Dept. of Health and Human Services, and protested that the researchers obtaining his family history without his explicit consent constituted a violation of his privacy rights. OHRP, apparently siding with the father, blocked the offspring from using the father's information and forbade any further attempts to obtain more information on grounds that an individual's (specifically his) right to privacy and autonomy is paramount.

Among the interesting and difficult issues in this case is the fact that it challenges us to think deeply about the weighted values we assign to first principles, namely the right to privacy. Whose right predominates in this case-the father's or the child's?

Laws exist to ensure medical privacy. New federal medical privacy rules under HIPPA spell out the requirements to ensure privacy of all individuals.

These rules, though scheduled to become practice February 2001, are facing strict opposition from several different sectors of the health care industry, primarily because of the cost and impracticalities involved in implementation. For numerous reasons, it is far from clear if these rules would support the father's claim, and if so how. In spite of our best efforts to anticipate and resolve ethical quandaries arising from the application of new genetic technologies, it is likely that unexpected conflicts will arise. Those discussed in this article are not intended as an exhaustive list.

Pharmacogenetics has fewer ethical issues than other medical biotechnologies.

The ethical issues here are remarkably similar to those standardly invoked in pre-dispositional testing discussions. Yet, arguably the stakes are lower here. The risk of psychological harm is, for the most part, far less substantial than testing for a late onset disorder like Huntington Disease, for which, effective treatment does not exist. Still, in the absence of guidance about what constitutes high and low stakes, ethically defensible decision making requires acknowledgement of the competing interests and a broad enough scope of concern to analyse how an apparent low risk can become a real high risk and vice versa.

Pharmacogenetics will ensure safer drugs.

The end product of this technology — individualized drugs — must be made easily available to all.

Pharmacogenetics will permit gene profiling to answer questions about medicine responses, as well as enable researchers to design better and safer medicines.

The science and its applications are real today and will be increasingly common in coming years. While the likelihood that individuals will be shut out from health insurance because they do not respond to a single drug or because a particular drug formula is toxic to them is extremely low, as would be employment exclusions (in hiring, promoting or job responsibilities), the issues underscore the importance of debating more widely the ethical use of pharmacogenetics.

Pharmacodynamics is the study of the biochemical and physiological effects of drugs on the body or on microorganisms or parasites within or on the body and the mechanisms of drug action and the relationship between drug concentration and effect. One dominant example is drug-receptor interactions as modelled by

$$L + R \rightleftharpoons L{\cdot}R$$

where

$L$ = ligand (drug),

$R$ = receptor (attachment site), reaction dynamics that can be studied mathematically through tools such as free energy maps.

Pharmacodynamics is often summarized as the study of what a drug does to the body, whereas pharmacokinetics is the study of what the body does to a drug. Pharmacodynamics is sometimes abbreviated as "PD", while pharmacokinetics can be referred to as "PK".

### Effects on the Body

The majority of drugs either (a) mimic or inhibit normal physiological/biochemical processes or inhibit pathological processes in animals or (b) inhibit vital processes of endo- or ectoparasites and microbial organisms.

There are 5 main drug actions:

- depressing
- stimulating
- destroying cells (cytotoxicity)
- irritation
- replacing substances.

### Desired activity

The desired activity of a drug is mainly due to one of the following:

- Cellular membrane disruption
- Chemical reaction
- Interaction with enzyme proteins
- Interaction with structural proteins
- Interaction with carrier proteins
- Interaction with ion channels.

Ligand binding to receptors:

- Hormone receptors
- Neuromodulator receptors
- Neurotransmitter receptors.

General anesthetics were once thought to work by disordering the neural membranes, thereby altering the $Na^+$ influx. Antacids and chelating agents combine chemically in the body. Enzyme-substrate binding is a way to alter the production or metabolism of key endogenous chemicals, for example aspirin irreversibly inhibits the enzyme prostaglandin synthetase (cyclooxygenase) thereby preventing inflammatory response. Colchicine, a drug for gout, interferes with the function of the structural protein tubulin, while Digitalis, a drug still used in heart failure, inhibits the activity of the carrier molecule, Na-K-ATPase pump. The widest class of drugs act as ligands which bind to receptors which determine cellular effects. Upon drug binding, receptors can elicit their normal action (agonist), blocked action (antagonist), or even action opposite to normal (inverse agonist).

In principle, a pharmacologist would aim for a target plasma concentration of the drug for a desired level of response. In reality, there are many factors affecting this goal. Pharmacokinetic factors determine peak concentrations, and concentrations cannot be maintained with absolute consistency because of metabolic breakdown and excretory clearance. Genetic factors may exist which would alter metabolism or drug action itself, and a patient's immediate status may also affect indicated dosage.

### Undesirable effects

Undesirable effects of a drug include:

- Increased probability of cell mutation (carcinogenic activity)

- A multitude of simultaneous assorted actions which may be deleterious
- Interaction (additive, multiplicative, or metabolic)
- Induced physiological damage, or abnormal chronic conditions.

### Therapeutic Window

The therapeutic window is the amount of a medication between the amount that gives an effect (effective dose) and the amount that gives more adverse effects than desired effects. For instance, medication with a small pharmaceutical window must be administered with care and control, e.g. by frequently measuring blood concentration of the drug, since it easily loses effects or gives adverse effects.

### Receptor Binding

The binding of ligands (drug) to receptors is governed by the *law of mass action* which relates the large-scale status to the rate of numerous molecular processes. The rates of formation and un-formation can be used to determine the equilibrium concentration of bound receptors. The *equilibrium dissociation constant* is defined by:

$$\mathrm{Kd} = \frac{[\mathrm{L}][\mathrm{R}]}{[\mathrm{L}\cdot\mathrm{R}]}.$$

$$\mathrm{L} + \mathrm{R} \leftrightarrow \mathrm{L}\cdot\mathrm{R}$$

where

$L$ = ligand,

$R$ = receptor, square brackets ] denote concentration.

The fraction of bound receptors is found as $(1+[R]/[L\cdot R])^{-1}$, which can then be expressed using $K_d$ as,

$$\text{Fraction Bound} = \frac{1}{1+\frac{K_d}{[L]}}$$

This expression is one way to consider the effect of a drug, in which the response is related to the fraction of bound receptors. The fraction of bound receptors is known as occupancy. The relationship between occupancy and pharmacological response is usually non-linear. This explains the so called *receptor reserve* phenomenon i.e. the concentration producing 50% occupancy is typically higher than the concentration producing 50% of maximum response.

Often the response is determined as a function of *log[L]* to consider many orders of magnitude of concentration. However, there is no biological or physical theory which relates effects to the log of concentration. It is just convenient for graphing purposes. It is useful to note that 50% of the receptors are bound when $[L]=K_d$.

The graph shown represents the conc-response for two hypothetical receptor agonists, plotted in a semi-log fashion. The curve toward the left represents a higher potency (potency arrow does not indicate direction of increase) since lower concentrations are needed for a given response. The effect increases as a function of concentration.

# Chapter 4

# Bioinformatics and Postgenomic Chemistry

Deciphering the human genome will probably identify 5,000 to 10,000 new proteins of therapeutic relevance; some will be homologous to existing known targets, but many if not most will be entirely novel, with unknown structure and function. Medicinal chemists know how to design and prepare molecules that can inhibit intracellular enzymes such as proteases, kinases, and phosphatases or modulate cell-surface or nuclear receptors with agonists or antagonists; indeed, many examples of such molecules have reached the drug market.

However, the arsenal of concepts, tools, technologies, and strategies that medicinal chemists use today are rather limited if the purpose is to control peptide-hormone/receptor interactions or protein-protein interactions. Mimicking the interaction of hormones, such as follicle-stimulating or growth hormones, with their receptors or controlling cytokines signalling/transcription pathways are examples of such challenges we face at Serono International.

New tools must be invented and put in place beyond high-throughput screening (HTS) and peptide-phase display technologies. Such tools will be at the frontier of synthetic, medicinal, analytical, and computational chemistry; cheminformatics; HTS technologies; structural biology; and cell biology. Success will depend on how efficiently these areas cross-fertilize each other. Thousands of new potential targets for drug discovery will be proposed by the genomics initiative. These numbers are huge compared with the pre-genomic era but are still very small when compared with the vastness of "chemical" space. Estimates of the number of possible "drug-type" small molecules—those with a molecular weight of less than 1,000—average around 1040.

The largest pharma companies today have collections of discrete compounds that do not exceed 10 million. In other words, the field of investigation for chemists remains essentially unexplored. Here, organic synthetic chemists will continue to play a major role by proposing new, original, sophisticated (3-D information-rich) yet readily accessible scaffolds or templates to be used in a combinatorial manner; inventing new reactions; developing new synthetic pathways; and extending their scope of application.

Such tools will be essential for medicinal chemists facing the post-genomic era. That's not all for chemists. One of the greatest challenges in the post-genomic era will be the validation of the most promising targets for therapeutic intervention. Chemists can play a critical role by providing small molecules that alter gene function by selectively binding to their protein products. Having a small molecule partner for every gene product is the dream; such a molecule would then be evaluated in cell models and in vivo disease models, thus validating the protein target for drug discovery. This process is called "chemical genetics" and allows chemists to be involved in the drug discovery process before the target is validated; previously, medicinal chemistry programs started only after target validation.

Chemical genetics is based on constructive interactions between chemists and biologists; if successful, it will have spectacular advantages compared to the classical validation tools based on proteins mutants and knockout organisms. Chemical genetics offers conditional techniques and, more importantly, allows concurrent analysis of both the importance of a protein in a disease process (target validation) and a protein's amenability to functional modulation by a small molecule (target tractability). Having a small molecule partner for every gene product represents a fantastic opportunity to design the next generation of drugs in phase with the pharmacogenetics vision of the drug market, where drug responses are assessed on the basis of genetic disposition.

Besides new target identification, genomics will also lead to sharp diagnostic tools; it's expected that by 2025, everybody can have her or his genome sequenced. We will therefore move from the era of "disease"-specific compounds to the era of "disease and patient"-specific compounds in which drugs will be chosen on the basis of the phenotype/genotype of the diseased cell or tissue and the genome of the patient.

So the challenge for the future is to discover the right drug for the right patient at the right dose in the right time. Chemists will play a major role in this tough challenge, not only by providing new tools and concepts in their own field but also as essential members of multidisciplinary teams. This is particularly true in biotech companies like Serono where chemists find the research and innovation culture that will make it possible.

## New Problems and Challenges

The past decade has been characterized by the dynamic integration of chemistry and biology. Chemistry as the science investigating structures and properties of molecules provides the fundamental basis for the development of biochemistry, molecular biology, and biotechnology. Chemical science actively investigates and employs the principles of molecular organization of biological processes in living systems. Chemistry and chemical technology peculiarly use biological materials and catalysts to create steadily increased variety of useful molecules. Genomic research also has a great impact on related fields of modern science, including chemistry.

The creation of a holistic system of knowledge of genetic nature, genetic materials, and structures of genomes has influence on the development of chemistry, providing it with evolutionary new representatives, methods, and molecular structures. In addition, the challenge of identification of chemical structure and function of all proteins in living organisms links chemical and biological science. The complete elucidation of the human genome and genomes of many other organisms is a remarkable scientific breakthrough that will affect our lives for decades to come.

The content of structural information generated as a result of genomics studies is huge. More than 500 genomes are already partially or completely sequenced. Genomes of various microorganisms, plants, animals, as well as the human genome, are among them. These studies allowed identification of 1.1 million nucleotide sequences that code for proteins.

However, there are certain limitations that need to be overcome before one can exploit successfully the wealth of information from genomic studies. Although not at the forefront during the past genomics efforts, chemistry will undeniably continue to play a crucial role in the biological sciences in the future. Therefore, one can predict a bright future for chemistry in the postgenomic era. For example, developments in physical, surface, and analytical chemistry will surely provide the basis of novel, improved detection methods that will speed up and facilitate diagnostics. Organic chemistry will influence biological sciences even more. Genomics provides chemistry with new possibilities, tasks, and challenges.

The functioning of a living organism is far more complex than the genome may suggest. Therefore, profound studies of proteome, metabolome, cellome, and physiome are under way. Another important feature is the multidisciplinarity of problems, which can be solved only by the combined and concerted efforts of scientific teams, including experts from different disciplines (chemists, biologists, physicians, mathematicians, and others).

There is a need to discuss the directions of future developments in various fields of chemistry in view of and with respect to the developments in genomics. Obviously, the achievements of genomics and proteomics could considerably influence the classical chemistry as well as result in new scientific directions (IUPAC Project # 2001-005-1-300 "Postgenomic chemistry").

## Identification of all Molecular Participants of Biological Processes

The fundamental and philosophical result of the genomic project is that all molecular representatives of the world of biological diversity can be identified.

### *Trace Element-containing Protein*

Selenoproteins (SPs) are responsible for most biomedical effects of dietary applicable selenium and are important for mammals because selenium plays a significant role in many physiological processes, namely, in male reproduction, aging, immune function, and cancer prevention. The information about the human SPs potentially can be used for the further systematic clarification of mammalian SPs and their functions. It was established that SPs contain selenium in the form of selenocysteine (Sec), which was named as the 21st amino acid.

There are three terminating codones in the universal genetic code. It was discovered that one of them (UGA) has dual function, signalizing termination of protein synthesis as well as incorporation of such amino acid as Sec. Since then, classical methods of biochemistry and a variety of biophysical techniques are not enough to analyse the correct function of UGA. Most known selenoprotein genes have homologues, in which Sec is replaced with cysteine (Cys). Using that, specific approach was suggested for identification of SPs. It implies the search for the Sec/Cys-containing protein pairs of homologs. In addition, to identify mammalian SPs, human and mouse (or human and rat) genomes were searched in parallel to identity pairs of Sec insertion sequence (SECIS) elements present in the selenoprotein genes. This method supported by the application of specially developed computer programs provided independent verification of a number of SPs in various genomes. Subsequent investigations of new SPs revealed examples of proteins with expression patterns limited to embryos or testes, as well as proteins with novel subcellular distributions. It was established that all characterized mammalian SPs are globular proteins. Sec is inserted into polypeptide chains during ribosome-based protein synthesis.

The expression of fusion SPs, containing a C-terminal green fluorescent protein (GFP) tag, and their following detection with antibodies to GFP by means of electron microscopy helped in revealing the first known plasma membrane SPs. It was found that the human genome has 25 selenoprotein genes and the mouse genome 24 ones. The further understanding of mechanisms of Sec insertion can assist in revealing approaches to the targeted insertion of this residue into proteins. These examples emphasize the power of bioinformatics in establishing of deep interrelation between genomics and chemistry.

Identification and characterization of new metal-containing proteins in humans and other organisms should set up new challenges and highlight the need for investigation of genomic and postgenomic chemistry of biological trace elements. Understanding of the identities and functions of trace elementcontaining proteins can also provide new tools for nonspecific incorporation of these elements into proteins.

## Phosphotriesterases in all known Genomes

In recent years, of notable interest to us has been the study of triesterases, the enzymes catalysing hydrolysis of triesters of phosphoric acid. The enzymes of this class can accelerate the hydrolysis of a wide group of phosphoric acid derivatives including such supertoxic compounds as Sarin, Soman, Vx, many organophosphorus pesticides, and toxins. The enzymes of this class are applied as detectors of toxic organophosphates in biosensors and considered as major components in new protective systems against weapons of mass destruction.

We offered the mechanism of catalysis including the steps of the electrophilic activation of substrate reaction site by a complex with one metallic ion (CoI) in enzyme-active site and of the electrophilic activation of water molecule by a complex with a second metallic ion (CoII). The application of modern computer methods for comparison of sequence of nucleic acids coding protein synthesis allows the researchers to answer the question of existence of homologous proteins possessing phosphoesterase activity. Twenty-five proteins detected were highly homologous with the beststudied enzyme isolated from *Flafobacterium* sp. cells. The procedure for construction of the phylogenetic tree includes the combination sequences in clusters with the follow-up comparison of proteins by the principle of minimal deviations. Structurally, the proteins unite in "branches"; the more similar the primary protein structures are, the closer they are located to one another on the phylogenetic tree. The necessary step in the active site formation and the catalytic activity manifestation is the presence of Lys169 in protein, the carbamylated form of which "collects" cobalt ions into a common active site and provides a necessary spatial closeness between substrate and water activation sites. It shows that, according to the presented data, human proteins and *E. coli* cells do not possess the organophosphate-related enzymatic activity.

Proteins of some microorganisms, such as *Mycobacterium, Sulfolobus, Mycoplasma*, and *Lystera*, show the potential enzymatic activity. This opens notable possibilities in search of new enzymes of this class.

## Chemical Proteomics: Management of Biosystems by Selective Ligands

The regulation of many processes taking place in the biological systems can be organized through application of effectors special for some key biomolecules. Thousands of unknown proteins have been identified through genomic and proteomic research. It is important to discover specific molecules that can interact with them selectively to understand the role of newly discovered proteins and to regulate their activity. The main goal is to find one low-molecular-weight compound able to block or stimulate the activity of one protein in a highly selective way. This area can be identified as "chemical proteomics". The goal of the chemical proteomics is the creation of "chemical keys" for all proteins existing in nature. Special interest to these tasks is dictated presumably by pharmacological importance.

More than 95 % of all human diseases are associated with defective gene expression or post-translation protein modification. Since that, specific molecular ligands can be very useful in explanation and restoration of the functions of biological macromolecules. They also can be used for construction of novel affinity tags to facilitate purification and detection of target proteins or polypeptides. The ligands helpful in control (suppression or activation) of gene expression have high importance. Specific binding between proteins and their ligands is the molecular basis of all essential biological processes. The ways used in identification and characterization of protein sites interacting with other molecules are based on different strategies. One of the modern suggested ways is the following. Selection of appropriate site on the target protein, design of a complementary ligand compatible with the 3D structure of the site, construction of a limited solid-phase combinatorial library of near-neighbor ligands and solution synthesis of the hit ligand, immobilization, optimization, and application of the adsorbent for the purification of the target protein. This strategy is directed to the understanding of affinity and selectivity for enzyme inhibitors uses information from ligands and 3D structures of target proteins.

Combinatorial chemistry has emerged as a set of novel strategies for the synthesis of large sets of compounds (combinatorial libraries) for biological evaluation. Combinatorial approaches capable of generating libraries of different proteins allow the directed screening for a particular biological activity.

Analysis of proteins in complex with their ligands provides structural information on the amino acid residues involved in ligand binding. Synthetic chemistry enables one to create a large number of different molecules, including simple and more complex peptides, which may contain different amino acids, including unnatural ones, as well as non-peptide small molecules. The design and generation of synthetic molecules that can interfere with protein–ligand interactions represents a promising strategy for the explanation of protein structure and function. In addition to their basic significance, such synthetic proteomimetics are also useful tools for a range of biomedical applications.

A new very fast developing approach to the investigation of protein–ligand interactions is aimed at synthesis of molecules, which, owing to their specific molecular structure, are capable of mimicking the binding sites of natural proteins. Various protein-binding sites are composed of parts that are remote in the amino acid sequence, but brought into proximity by protein folding. They can be mimicked through scaffold peptides, in which the peptide fragments making up the binding site are presented through a molecular scaffold in a nonlinear, discontinuous fashion.

This task was initially accomplished by using short synthetic peptides to establish the linear binding sites of proteins. Such peptides, representing overlapping fragments of the protein, were tested individually for binding to the respective ligand in order to identify the protein region(s) responsible for the ligand recognition. The epitopes identified by this way were characterized and optimized regarding their binding affinity to the respective ligands.

Synthetic chemistry should play a significant role in the identification and functional characterization of gene products, as well as in the investigation of proteins' functions through controlled interference with the protein–ligand interactions.

### *Biocatalysis: Enzymes of New Generation*

The methods of bioinformatics allow identification of comprehensive sets of genes encoding enzymes in various organisms as well as comparative and functional analysis of these proteins. For example, the problem of enzyme thermostability could be solved at the genetic level by comparing genomes of mesophilic and thermophilic microorganisms. Genomic access to thermostable enzymes and proteins may result in development of unique technologies of high-temperature biotechnological processes. Enzymes are the most widespread catalysts that can be obtained from renewable raw materials.

Biocatalysis is the basis of important chemical processes. During the past two centuries, organic chemistry was substantially oriented toward the transformation of hydrocarbons and their different chemical derivatives, whereas the chemistry of the 21st century will be dealing with the problems of chemical transformation of renewable raw materials such as carbohydrates, biomass components, and carbon dioxide. Biocatalysts are the best tools for the chemical modification of molecules in the 21st century. The detailed structural information at the atomic level is already known for a large variety of enzymes from X-ray crystallographic and NMR spectrometric methods. The existing concepts explain the observed effects of acceleration of chemical reactions and the nature of enzyme specificity and are based on fundamental physicochemical laws.

During the last 50 years, enzymes and biocatalytic systems were realized in the wide range of applications, including fine organic synthesis and multiple forms of chemical analysis. Enzymes provided the basis of many processes in food industry, industrial production, and utilization of detergents. They served as key components in environmental biotechnology and medicine and improved chemical and biological safety of various industrial processes.

Many "weaknesses" of enzymatic catalysis were overcome during the past decades. Thus, the methods of effective stabilization of enzymes were developed and the new conditions for their exploitation were discovered, and new methods of the use of enzymes in the media with organic solvents at increased temperatures were described. Methods of genetic engineering and site-directed mutagenesis played a significant role in elucidating reaction mechanisms of various enzymes and providing molecular basis for their biotechnological applications. Potentially all these features make enzymes the most effective biocatalysts. The up-to-date prognosis suggests that within one decade, total enzyme production in the world will reach the level of or even exceed the production of "classic" catalysts used in the chemical industry within the next decade. The creation and construction of enzymes with new properties is based on the application of genetic information and modern achievements of genomics. It is possible to expect the appearance of new scientific directions of enzyme investigations and the development of new biocatalytical technologies.

The use of new methods of preparation of gene-expressed proteins obtained from unnatural amino acids might be capable of supporting the development of principally new catalysts, considerably widening the possibilities of biocatalysis. Biosynthesis of enzymes with unnatural amino acids included in the polymer chain could result in the development of new types of active sites, solving the problem of the use of enzymatic reactions under extreme conditions (extreme pH, temperature, salt content, etc.). The chemical modification of active sites by unnatural amino acids should result in the application of enzyme families with altered catalytic efficiencies and reaction mechanisms, and those that are characterized by transformed specificity and enantioselectivity.

In addition to methods traditionally used in protein design, such as DNA-shuffling, and directed evolution, it is anticipated that new methods allowing generation of enzymes with predefined properties will emerge. In this regard, the new methods allowing construction of proteins from the limited number of essential amino acids are quite attractive. These approaches should result in deeper understanding of physiological functions of proteins and lead to the appearance of new biocatalyst families of high applied significance.

## Biocatalysis: New Resource Technologies

The limitations of traditional sources used in the chemical industry as well as limitations of hydrocarbon sources of energy, illustrate the need for development of new raw and energy sources. Growing interest in the biocatalytic conversion of biomass as the initial chemical raw materials for production of new materials, polymers, and energy sources (i.e., hydrogen, methane, ethanol, and diesel oil) may be envisioned as characteristic of the next decades. Significant progress may be expected in the development of biosystems catalysing the reduction of carbon dioxide to the basic carbon sources for the chemical industry (organic acids, alcohols, and monomers).

### Enzyme Catalytic Sites and Mechanism of Catalysis from Sequence Data

Computer methods are widely employed in modern chemistry and molecular biology for various purposes. During the last decade, two approaches, bioinformatics and molecular modelling, have become especially popular and intensively developing areas. The methodology of bioinformatics is based on the informational analysis of nucleotide sequences and proteins. Within a framework of adequate physical considerations, molecular modelling allows to characterize protein structures and their changes induced by some treatments or local structural alterations. In combination with methods of molecular dynamics, molecular modelling represents a powerful approach that provides understanding of numerous physical and chemical aspects of protein molecules. Graphic capacities of modern work stations, relatively low cost and potency of modern processors, storage capacity of accumulated information, availability of a large number of resolved spatial protein structures, and nucleic acids and their complexes are the background for rapid worldwide distribution of computer modelling methods among scientists. These methods of computer modelling are widely used for determination of structures of biological molecules (X-ray analysis and multimer nuclear magnetic resonance), theoretical studies of their interactions, studies of the interaction between membrane and membrane proteins, structural analysis of site-directed mutagenesis and prediction of new mutation sites, and protein structure prediction (method of homologous modelling).

Well-known software allow the construction of various biological molecules, edit their structure (by substituting some groups, changing conformation of molecular groups) followed by subsequent structural relaxation by means of methods of molecular mechanics and dynamics. These programs also allow the calculation of various characteristics of the constructed molecules (e.g., electrostatic potential, solvent accessible surface, etc.), study of interactions between various molecules, alignment of protein sequences, creation of homology-based structures, and the investigation of ligand docking to active sites of enzymes.

### Multiple Alignment of Amino Acid Sequences allows Recognition of the Enzyme Catalytic Site

Amino acid sequence determines structure and properties of each protein. Now, good evidence exists that in almost endless variability of proteins some structural elements are rather conservative, and these elements mainly determine the function of the protein molecule. This is especially demonstrative in the case of catalytic proteins. For example, in the case of hydrolases, which represent about one-third of all enzymes (about 1100 of 3700 enzymes) listed in the enzyme classification, only four main types of sites forming the catalytic structure are known. Consideration of active site structure of enzymes requires subdivision of the active site into two structural constituents:

- substrate-binding subsite, which is responsible for binding, fixation, and certain orientation of substrate(s); it determines enzyme specificity;
- catalytic subsite, which is responsible for chemical transformation of substrate molecule; this site usually employs general acid-base catalysis.

It is possible that within one large enzyme superfamily the substrate-binding subsite responsible for the enzyme specificity exists as quite variable protein structure corresponding to variations in the substrate structures. However, catalytic sites should represent rather conservative structural elements due to a limited number of catalytic site types. To test this hypothesis, we have employed the bioinformatic approach based on the comparison of amino acid sequences of proteins constituting one large family. We analysed results of sequence alignment of a few large enzyme families from HSSP database. These enzyme families were selected by the following criteria:

- the number of analysed enzymes should exceed 100; this provides reasonable statistical significance of the results;
- this analysis requires selection of enzymes with known active site structures and well documented catalytic mechanism.

One of the quantitative criteria of position conservatism for each residue in the protein sequence is the statistical criterion in the form of Shannon entropy. In information theory, Shannon entropy is one of the most important functions. The informational entropy (Shannon entropy) is a very convenient function for comparison of related proteins with distinct amino acid sequences. The alignment procedure vs. some reference protein represents sequence positioning one over another one, followed by fixation of homologous sites, and recognition and eliminations of inserts. Such comparison of a large number of protein sequences allows the calculation of probability of localization of some amino acid residue in a certain position. This probability is determined as relative frequency of the amino acid $j$ in a given position $i$.

Hydrolases were used as the object of research. Four mechanisms of water activation in reactions catalysed by these enzymes are known. They involve:

- aspartic (glutamic) acid carboxyl group, water activation by nucleophilic mechanism;
- histidine imidazole group, water activation by nucleophilic mechanism;
- complex with zinc or cobalt ions, water activation by electrophilic mechanism;
- complex with magnesium or manganese ions, water activation by electrophilic mechanism.

The following enzyme families representing these four mechanisms of water activation were chosen: pepsin family (activation by carboxyl group), chymotrypsin and subtilisin family (activation by imidazole group), alkaline phosphatase family (activation by complex with zinc or cobalt ions), pyrophosphatase family (activation by complex with magnesium or manganese ion).

Analysis of highly conservative amino acids (for which $Hj \sim 0$) revealed that during alignment, amino acid residues forming catalytically active subsites always represent conservative elements of the protein sequence. The catalytic site of acidic pepsin type proteases includes carboxyl groups of Asp31 and Asp215. These residues are recognized during alignment of amino acid sequences of pepsin family as conservative positions characterized by minimal value of Shannon entropy. The catalytic site of the chymotrypsin family includes Ser195, His57, and Asp102, whereas the catalytic site of the alkaline phosphatase family includes Asp51, Asp369, His370, Asp327, His412, His331, and Ser102. Catalysis by inorganic phosphatase involves Glu20, Asp65, Asp70, and Asp102.

All these residues are recognized as conservative ones during multiple alignment of amino acid sequences of corresponding protein families. Thus, the bioinformatic approach allows recognition of side chains of amino acid residues forming a catalytic site of the enzymes and responsible for nucleophilic/electrophilic substrate conversion. The comparison of enzyme amino acid sequences also revealed that Gly and Asp are the most frequently recognized as absolutely conservative residues. The finding that Gly is the most conservative residue was rather unexpected. Asp takes the second position in this list, and the sum of Gly and Asp represents about 50 % of all conservative residues recognized.

Amino acid residues were ranked by their conservatism in these four protein families. For each amino acid, we determined its frequency as the conservative element ($Hj \sim 0$), normalization on total number of conservative positions for all amino acid residues in these families.

Total frequency of conservative residues in enzymes completely differs from total frequency of amino acid residues in proteins, where Leu is the most frequent residues, whereas Gly takes only the fourth place in this list. (Frequency of Gly as the most conservative residue is 37 %.) Gly, Asp, Cys, Pro, and His are the most frequent conservative residues in enzymes. They represent about 70 % of all conservative positions in these enzymes, whereas Met and Ile represent the most variable elements in the amino acid sequences. Thus, the most conservative residues can be separated into two principally different groups: (1) residues involved in substrate activation and acting as acids and bases (Asp and His); (2) residues forming active site architecture (Gly, Cys, Pro).

### *Aspartic Acid and Histidine in the Enzymatic Catalytic Cycle*

The bioinformatic approach used in this study demonstrates a crucial role of aspartic acid and histidine residues in the functioning of the active site. Let us consider this role in more detail based on the analysis of mechanism of hydrolase action. As mentioned above, hydrolases represent the largest class of all known enzymes and their molecular mechanisms of catalysis have been well characterized. For most hydrolases, functional groups of amino acid residues constituting catalytic sites have also been identified and the interaction between these groups during catalytic cycle has been elucidated.

Based on the active site structure and mechanism of action all hydrolases can be arbitrary subdivided into the four main types:

- hydrolases containing aspartic or glutamic acid residues in the active site (lysozyme-pepsin type);
- hydrolases using imidazole group for water activation (the type of pancreatic ribonuclease, chymotrypsin, subtilisin, papain, lipase);
- hydrolases using complexes with Zn2+ or Co2+ for activation of water and substrate (type of alkaline phosphatase, carboxypeptidase A, and organophosphate hydrolase);
- hydrolases using Mg2+ and Mn2+ for activation of water and substrate (inorganic pyrophosphatase type).

Analysis of the catalytic mechanisms revealed that in most types of catalytic sites aspartic acid and histidine play principally important role.

### *Role of Gly in Formation and Conformational Flexibility of Active Structure*

It is clear that conservative Gly residues do not play an important role in the activation of water molecules during the catalytic cycle because Gly does not have any substituent at the *alpha*-carbon atom, and therefore it lacks pronounced chemical function.

Nevertheless, Gly residues are essential for protein structure. This was demonstrated by mutation experiments where conservative Gly residues were substituted for any other amino acid. As a rule such substitutions resulted either in complete loss or significant decrease in catalytic activity. Apparently, conservative Gly residues are principally important for the following functions.

Being a unique amino acid with the most energetically favorable rotation along C–N and C–C bonds of polypeptide chain (Õ and ø Ramachandran angles) glycine may be a "junction point", providing change in the direction of polypeptide chain during "assembly" of amino acid residues into functionally competent active site. The essential role of Gly for structural function of enzyme-active sites was demonstrated by molecular modelling method. Thus, the presence of conservative glycines may explain the structural paradox of enzymatic catalysis, when completely identical active sites are formed from completely distinct protein sequences. They just share several common features such as the existence of conservative glycines and factors stabilizing the assembled structure. In this connection, we should mention that cysteine residues involved in disulfide bond formation (one of the most common stabilizing factors) take the third place in the rating of conservatism.

Conservative glycine residues may also function as a "hinge" responsible for known conformational flexibility of the active site. In many cases, conservative glycines are located near catalytically active residues. For example, the following conservative motifs were found in hydrolases from various families: Asp215XGly217 (pepsin), Asp170XXGly173 (thermolysin), Asp32XGly, His63Gly64, Gly119XSer221 (subtilisin), Gly17XSer177 (trypsin), His76Gly77, Ser153XGly155, Gly75XAsp177 (lipases). In these enzymes, Asp, Ser, and His residues are from the active site structures.

For some enzymes, values of *phi* and *iota* angles for amino acid residues of the catalytic site are out of the energetically "relaxed limits". This was found for amino acid residues of *alpha*-chymotrypsin (His57, Asp102, Ser195) by using the Ramachandran map of these residues.

The active site of these enzymes is conformationally tensed (values of Õ and ø angles are in an energetically unfavorable region).

In enzymatic catalysis, the conversion of initial substrate into reaction product(s) involves a series of intermediates possessing distinct structures. Conservative glycines of the active site may function as "relaxing" elements adapting active site conformation for the catalytic conversion of the next intermediate. Cysteine and proline residues take top positions in the rating of amino acid conservatism. They are also essential for active site formation. Proline is a unique amino acid that unfolds the polypeptide chain.

Cysteine residues provide "required" active site structure by forming a disulfide bridge, which promotes fixation of catalytic residues (often located in different positions of a polypeptide chain) in the active site structure. For many enzymes, disulfide bond formation is the final step terminating active site assembly.

### *Enzymes vs. Classical Catalysts*

A wide diversity of enzymes, great potentials of enzyme modification by genetic engineering methods, and a deep insight into enzyme-related mechanisms potentially provide their diverse applications. Enzymes are widely applied in fine organic synthesis, medicine, and analysis as active components of washing powders. Enzymes are most widely applied in the food industry. However, enzymes can also be fairly competitive in solving some classical chemical tasks. As example, consider the problem of creation of catalysts for fuel cells. In view of progressing deficit of hydrocarbons, as a basis of modern power engineering, much attention is nowadays attracted to fuel cells, i.e., the electrochemical energy converters where fuel oxidation is spatially separated from reduction of oxidizer, usually air oxygen, resulting in the electrochemical potential generated on two electrodes.

Fuel cells afford a direct conversion of oxidation energy to the electric form without the intermediate emission of heat. Due to this fact, the efficiency of fuel cells is fairly high, approaching 100 %. The electrocatalytic problem of fuel oxidation and oxygen reduction is quite involved. On the one hand, the electrocatalyst should accelerate the electron transfer from the electrode matrix to oxygen. On the other hand, it should take an electron out of the fuel molecule without intermediate formation of free-radical particles in both anodic and cathodic reactions. At present, hydrogen–oxygen fuel cells are most elaborated. Supposedly, the fuel cells are an alternative to the internal combustion engines for motorcars. The only appropriate catalyst in catalytic and corrosive parameters for hydrogen–oxygen cells is metallic platinum. However, platinum, as electrocatalyst for fuel cells, has some principal drawbacks: first of all, the high price and sensitivity and, hence, the quantity of CO and H2S in the fuel. CO actually totally and irreversibly inhibits platinum at 100 ppm concentration of the former in the fuel hydrogen. A principal limitation of platinum as catalysts in fuel cells for motorcars is imposed by highly limited resources for the world production of platinum. To-date, the world production is about 180 t platinum per year. To produce 60 million motorcars per year (the level of motorcar production in 2000), about 5800 t platinum are required according to the most modest estimates.

Biological catalysts are totally renewable and industrially producible in any necessary quantity.

To create the enzyme-based fuel cells, we applied hydrogenases (hydrogen oxidation enzymes) and blue copper-containing oxidases of laccase type (oxygen reduction enzyme). The fundamental ground for creation of enzyme fuel cells is the application of biocatalysis, i.e., the enzyme-catalysed acceleration of electrochemical reactions of electron transfer at the interface electronic conductor/ionic conductor upon direct "electric catalysis" between conductor and enzyme-active site.

The effect of direct electron transfer between conductor and enzyme-active site was disclosed in our laboratory and registered as the discovery (State registration No. 311, 1987). It should be noted that the USSR had the system of registration of priority research "discoveries", which did not follow from direct theoretical investigations, were unexpected, and promised novel technological advances. It was shown that at the electrodes made of any electron-conducting material, hydrogenase chemosorption in air oxygen establishes a reversible equilibrium of hydrogen potential.

At the electrodes with sorbed laccase, the potential is established that approached the equilibrium potential of oxygen reduction. Hydrogen and oxygen enzyme electrodes underlay the construction of the fuel cell, affording us to obtain a record coefficient of conversion for the free energy of hydrogen oxidation by oxygen, 95.7 %. The latest information on enzymatic hydrogen cells was presented elsewhere. The detailed comparison of electrochemical behaviour of platinum and hydrogenase in hydrogen oxidation reaction showed that the exchange currents for hydrogenase by 2–3 factors higher than for platinum (referred to the catalyst active site). Note that the hydrogenase electrodes are not inhibited by H2S and practically not inhibited by

CO. Thus, for hydrogen activation reaction in electrochemical mode, it turned out that the enzymes have notably higher parameters than the "classical" catalyst. The possibility to produce enzymes in any necessary quantities is the ground for real practical creation of fuel cells with unique technological parameters as energy converters.

## Chemically Modified Proteins: New Properties, New Capabilities

Structural and functional proteomics has already identified a large number of proteins with novel folds designed to carry out specific biochemical reactions. Some of these proteins have inspired chemists and biochemists to engineer artificial and semi-artificial proteins, which may be used in unprecedented industrial or pharmacological applications. Therefore, the preparation of "synthetic" proteins with improved properties can become a rapidly developing part of postgenomic chemistry.

### *Fluoro-containing Proteins and Organisms*

The interest to the unnatural amino acid-containing proteins is stimulated by at least three aspects.

These are: (1) A potential possibility of the unnatural amino acid incorporation into proteins infinitely widens the spectrum of the novel proteins to be synthesized. The elaboration of the effective methods to synthesize the proteins and enzymes containing element-organic amino acid analogs gives us hope to obtain the proteins and enzymes with the properties previously unknown; (2) Novel classes of anticancer and antiviral agents could be proposed using the amino acid analogs and their derivatives; (3) The replacement of any amino acid significant part by their synthetic analogs is expected to lead to the formation of microorganisms with qualitatively novel properties. Amino-acyl tRNA synthetase is the key enzyme of amino acid recognition in the protein biosynthesis. Some analogs could not be incorporated into the proteins because of the high specificity of the above-mentioned synthetase.

Nevertheless, there are several main ways of unnatural amino acids' incorporation in protein structure. One way represents itself: cultivation amino acid auxotrophic *E. coli* strain with a cloned enzyme gene under the inducible promoter or in the presence of a specified protein expression inhibitor in a medium containing all natural amino acids. After significant biomass amount is accumulated, the medium is replaced and an amino acid analog is introduced in place of its natural counterpart. After a short incubation period, the protein expression is induced by the addition of inducer. The second way supposes chemical charging of the suppressor amino-acyl tRNA with the amino acid analog and an in vitro introduction of the analog at the nonsense mutated positions of the enzyme gene under investigation in the cell-free translation system. Another way to overcome the specificity of amino-acyl tRNA synthetase is mutagenesis of the enzyme. This approach may widen the enzyme specificity and enables one to incorporate the substances from a long analog list.

Realizing discussed approaches, the Tyr-auxotrophic strain *E. coli* B-2935 was transformed with plasmid pTES-His-OPH and inoculated in the medium containing F-Tyr. The biosynthesis of F-labeled organophosphate hydrolase was carried out. The mass-spectrometric analysis (MALDI-TOF) of purified protein testified to the replacement of 80 % Tyr-residues by their F-containing analogs (data not published).

It was shown that phenylalanine analog, namely, 4-fluorophenylalanine, can be introduced into the cellular protein by biosynthetic machinery. It was revealed that the relative content of FPA in the whole yeast protein from the cells grown up in the medium contained 1 mg/ml FPA.

### Hybrid Proteins

The use of recombinant proteins has increased significantly in recent years as a response to the rapidly developed fields of proteomics. Recombinant hybrids containing a polypeptide fusion partner, termed affinity tag, to facilitate the purification of the target polypeptides are widely used. Many different proteins, domains, or peptides can be fused with the target protein. The production of recombinant proteins in a highly purified and well-characterized form has become a major task for the protein chemistry in the postgenomic era. Several different strategies have been developed to produce recombinant proteins on a large scale. One approach is to use a very small peptide tag that should not interfere with the fused protein properties. The effect on tertiary structure and biological activity of fusion proteins with small tags depends on the location and on the amino acid composition of the tag. Another approach is to use large peptides or proteins as the fusion partner since the large partners can increase the stability and solubility of the target proteins. Some hybrid partners are specially used for the investigation of protein–protein interactions or changing properties of a target protein. For example, the catalytic properties of organophosphate hydrolase containing oligohistidine sequence considerably differed from the native enzyme. The optimum pH for hybrid enzyme shifted to the basic area and became equal to 10.5 instead of 9.0 that was typical for the native enzyme. The thermostability of the developed enzyme increased, and temperature optimum of action shifted from 48 °C up to 55 °C. In addition to that, the effective constants for hydrolysis of many substances catalysed by the hybrid enzyme enhanced 10 times compared to the native form.

The wide application of green fluorescent protein as fusion partner opened new capabilities in the field of proteomics. The importance of affinity-tag technology will increase because of its necessity for the peptide/protein chip design, high-throughput purification, peptide/ protein libraries, drug delivery systems, and strategies of large-scale production.

### *Postgenomic Macromolecular Chemistry*

Future technology will be based on materials with excellent performance and novel functions. There is a large demand for so-called intelligent smart materials: self-recovery, self-adjustment or control, selfdiagnosis, stand-by capability for detecting nonlinear onset, ability to be externally tuned, etc.

Recombinant DNA technology can be utilized to produce polymers with specific properties and behavioural characteristics. The production of polymers by recombinant methods affords significant advantages over traditional polymer synthesis technologies, including control of microstructure, stereochemistry, and biocompatibility.

The production of artificial proteins by recombinant methods has already been demonstrated in several laboratories. Some groups have succeeded in the development of artificial proteins with smart properties. The electroactive polymers have been used to make artificial muscles. Hybrid intelligent hydrogels were developed via self-assembly of synthetic polymers and recombinant polypeptides.

Modern gene technology allows expression of any desirable amino acid sequence, and the recent breakthrough in industrial microbial cell fermentation and downstream processing of proteins allows production of large quantities of recombinant polymers. Precise control of the composition and structure of recombinant polymers renders them with unique properties. The introduction of unnatural amino acids in the structure of recombinant polymers will provide them with further unique properties not observed in natural polymers.

Recombinant polymers will combine both high mechanical and chemical stability as construction materials with the ability to react in a strong and predictable way to environmental changes and, hence, perform as active materials. In fact, these polymers will pave a way to soft machines where chemical energy is converted to mechanical energy with high efficiency utilizing the conformational transitions of polymers rather then movement of solid machinery. The development and production of recombinant polymers is also a response to the demand for "green" alternatives to petroleum-based chemistry and waste-generating stoichiometric reagents. The recombinant polymers are produced by the environmentally friendly process of fermentation of microorganisms.

The resources needed for the production of recombinant polymers are renewable, and byproducts generated (e.g., microbial biomass) could be easily utilized and converted. Moreover, the recombinant polymers are biodegradable and pose minimal environmental threat as compared to traditional synthetic polymers. One could easily foresee that recombinant polymers along with other natural polymers like carbohydrates, polyesters, and different composite materials based on these polymers, would to a large extent replace many traditional synthetic polymers. Some of natural polymers are synthesized using template-guided synthesis. A classical example of this type of the synthesis is polymerase-catalysed replication of DNA molecules where novel chain is produced by one of two chains of primary DNA. In this case, it allows production of biomolecules identical to the template.

One could envision the development of this approach into several directions:

- development and investigation of chemical models of DNA polymerase; fine-tuning of conditions for the DNA proliferation in the systems in vitro with the use of chemical analogs of DNA copolymers;
- widening of possible application of enzymes (polymerases, oxidoreductases, esterases) for the template-directed matrix synthesis (polyaniline, polyethers, etc.); and
- investigation of catalytic synthesis of polymers that employ proliferation of polymer matrix; chemical models of polymerase chain reaction.

Investigations in this field should result in the creation of self-proliferating polymers. The new systems recording the information in chemical language different from the classical biological one could be suggested on the basis of a number of such polymers.

### New Bioanalytical Chemistry

A number of new analytical technologies that qualitatively changed the face of modern bioanalysis have been developed in various genomic projects and their further applications. In addition to DNA sequencing methods, considerable progress was achieved in the development of biochip analytical devices and biosensors. The development of proteomics illustrates the application of mass-spectrometry method for the analysis of proteins and peptides.

Advances in genomics, in combination with the progress in nano/materials/info technologies, are being integrated to enable analytical devices and systems with potential global effects on individual and public health, safety, economic, social and political systems.

DNA microarrays have revolutionized the way of nucleic acid characterization. One can easily foresee that protein and small-molecule microarrays will have the same impact in many areas, such as protein structure–function and protein–protein interaction studies, as well as in lead compound identification. Toward such practical microarrays, novel chemistries that ensure selective, reliable, and efficient linkages of synthetic and biological molecules onto solid support are desperately needed. It is reasonable to believe that new reagents and chemoselective reactions operating in water will significantly improve the preparation of useful, stable microarrays that will find applications in many areas, especially in the drug development process.

General principles of DNA polymerase reactions and PCR technique initiated development of a number of atypical chemical processes. It was the first example of single-molecule registration.

A single enzyme molecule can generate polymerization and formation of nanoparticles that can play the roles of markers of immunological interactions and nucleic acid-based reactions. Scanning probe microscopy can be used as a counter of such nanoparticles. This approach promises to be extremely specific at registration of single molecular interactions.

## Genome Projects of Micro-organisms

### *Biotechnology*

Biotechnology is a field of applied biology that involves the use of living organisms and bioprocesses in engineering, technology, medicine and other fields requiring bioproducts. Modern use similar term includes genetic engineering as well as cell-and tissue culture technologies. The concept encompasses a wide range of procedures (and history) for modifying living organisms according to human purposes-going back to domestication of animals, cultivation of plants, and "improvements" to these through breeding programs that employ artificial selection and hybridization. By comparison to biotechnology, bioengineering is generally thought of as a related field with its emphasis more on higher systems approaches (not necessarily altering or using biological materials *directly*) for interfacing with and utilizing living things. The United Nations Convention on Biological Diversity defines biotechnology as:

> *"Any technological application that uses biological systems, living organisms, or derivatives thereof, to make or modify products or processes for specific use."*

Biotechnology draws on the pure biological sciences (genetics, microbiology, animal cell culture, molecular biology, biochemistry, embryology, cell biology) and in many instances is also dependent on knowledge and methods from outside the sphere of biology (chemical engineering, bioprocess engineering, information technology, biorobotics). Conversely, modern biological sciences (including even concepts such as molecular ecology) are intimately entwined and dependent on the methods developed through biotechnology and what is commonly thought of as the life sciences industry.

### History

Biotechnology is not limited to medical/health applications (*unlike* Biomedical Engineering, which includes much biotechnology). Although not normally thought of as biotechnology, agriculture clearly fits the broad definition of "*using a biotechnological system to make products*" such that the cultivation of plants may be viewed as the earliest biotechnological enterprise. Agriculture has been theorized to have become the dominant way of producing food since the Neolithic Revolution. The processes and methods of agriculture have been refined by other mechanical and biological sciences since its inception. Through early biotechnology, farmers were able to select the best suited crops, having the highest yields, to produce enough food to support a growing population. Other uses of biotechnology were required as the crops and fields became increasingly large and difficult to maintain. Specific organisms and organism by-products were used to fertilize, restore nitrogen, and control pests.

Throughout the use of agriculture, farmers have inadvertently altered the genetics of their crops through introducing them to new environments and breeding them with other plants—one of the first forms of biotechnology. Cultures such as those in Mesopotamia, Egypt, and India developed the process of brewing beer.

It is still done by the same basic method of using malted grains (containing enzymes) to convert starch from grains into sugar and then adding specific yeasts to produce beer. In this process the carbohydrates in the grains were broken down into alcohols such as ethanol. Ancient Indians also used the juices of the plant Ephedra vulgaris and used to call it Soma. Later other cultures produced the process of Lactic acid fermentation which allowed the fermentation and preservation of other forms of food. Fermentation was also used in this time period to produce leavened bread. Although the process of fermentation was not fully understood until Pasteur's work in 1857, it is still the first use of biotechnology to convert a food source into another form.

For thousands of years, humans have used selective breeding to improve production of crops and livestock to use them for food. In selective breeding, organisms with desirable characteristics are mated to produce offspring with the same characteristics. For example, this technique was used with corn to produce the largest and sweetest crops.

In the early twentieth century scientists gained a greater understanding of microbiology and explored ways of manufacturing specific products. In 1917, Chaim Weizmann first used a pure microbiological culture in an industrial process, that of manufacturing corn starch using *Clostridium acetobutylicum,* to produce acetone, which the United Kingdom desperately needed to manufacture explosives during World War I.

Biotechnology has led to the development of antibiotics. In 1928, Alexander Fleming discovered the mold Penicillium. His work led to the purification of the antibiotic by Howard Florey, Ernst Boris Chain and Norman Heatley penicillin. In 1940, penicillin became available for medicinal use to treat bacterial infections in humans.

The field of modern biotechnology is thought to have largely begun on June 16, 1980, when the United States Supreme Court ruled that a genetically modified microorganism could be patented in the case of *Diamond v. Chakrabarty*. Indian-born Ananda Chakrabarty, working for General Electric, had developed a bacterium (derived from the *Pseudomonas* genus) capable of breaking down crude oil, which he proposed to use in treating oil spills.

Revenue in the industry is expected to grow by 12.9% in 2008. Another factor influencing the biotechnology sector's success is improved intellectual property rights legislation—and enforcement—worldwide, as well as strengthened demand for medical and pharmaceutical products to cope with an ageing, and ailing, U.S. population.

Rising demand for biofuels is expected to be good news for the biotechnology sector, with the Department of Energy estimating ethanol usage could reduce U.S. petroleum-derived fuel consumption by up to 30% by 2030. The biotechnology sector has allowed the U.S. farming industry to rapidly increase its supply of corn and soybeans—the main inputs into biofuels—by developing genetically modified seeds which are resistant to pests and drought. By boosting farm productivity, biotechnology plays a crucial role in ensuring that biofuel production targets are met.

## Applications

Biotechnology has applications in four major industrial areas, including health care (medical), crop production and agriculture, non food (industrial) uses of crops and other products (e.g. biodegradable plastics, vegetable oil, biofuels), and environmental uses.

For example, one application of biotechnology is the directed use of organisms for the manufacture of organic products (examples include beer and milk products). Another example is using naturally present bacteria by the mining industry in bioleaching. Biotechnology is also used to recycle, treat waste, clean up sites contaminated by industrial activities (bioremediation), and also to produce biological weapons. A series of derived terms have been coined to identify several branches of biotechnology, for example:

- Bioinformatics is an interdisciplinary field which addresses biological problems using computational techniques, and makes the rapid organization and analysis of biological data possible. The field may also be referred to as *computational biology*, and can be defined as, "conceptualizing biology in terms of molecules and then applying informatics techniques to understand and organize the information associated with these molecules, on a large scale." Bioinformatics plays a key role in various areas, such as functional genomics, structural genomics, and proteomics, and forms a key component in the biotechnology and pharmaceutical sector.
- Blue biotechnology is a term that has been used to describe the marine and aquatic applications of biotechnology, but its use is relatively rare.
- Green biotechnology is biotechnology applied to agricultural processes. An example would be the selection and domestication of plants via micropropagation. Another example is the designing of transgenic plants to grow under specific environments in the presence (or absence) of chemicals. One hope is that green biotechnology might produce more environmentally friendly

solutions than traditional industrial agriculture. An example of this is the engineering of a plant to express a pesticide, thereby ending the need of external application of pesticides. An example of this would be Bt corn. Whether or not green biotechnology products such as this are ultimately more environmentally friendly is a topic of considerable debate.

- Red biotechnology is applied to medical processes. Some examples are the designing of organisms to produce antibiotics, and the engineering of genetic cures through genetic manipulation.
- White biotechnology, also known as industrial biotechnology, is biotechnology applied to industrial processes. An example is the designing of an organism to produce a useful chemical. Another example is the using of enzymes as industrial catalysts to either produce valuable chemicals or destroy hazardous/ polluting chemicals. White biotechnology tends to consume less in resources than traditional processes used to produce industrial goods. The investment and economic output of all of these types of applied biotechnologies is termed as bioeconomy.

### Medicine

In medicine, modern biotechnology finds promising applications in such areas as:

- drug production
- pharmacogenomics
- gene therapy
- genetic testing: techniques in molecular biology detect genetic diseases. To test the developing fetus for Down syndrome, Amniocentesis and chorionic villus sampling can be used.

### Pharmacogenomics

Pharmacogenomics is the study of how the genetic inheritance of an individual affects his/her body's response to drugs. It is a coined word derived from the words "pharmacology" and "genomics". It is hence the study of the relationship between pharmaceuticals and genetics. The vision of pharmacogenomics is to be able to design and produce drugs that are adapted to each person's genetic makeup.

Pharmacogenomics results in the following benefits:

1. Development of tailor-made medicines. Using pharmacogenomics, pharmaceutical companies can create drugs based on the proteins, enzymes and RNA molecules that are associated with

specific genes and diseases. These tailor-made drugs promise not only to maximize therapeutic effects but also to decrease damage to nearby healthy cells.

2. More accurate methods of determining appropriate drug dosages. Knowing a patient's genetics will enable doctors to determine how well his/her body can process and metabolize a medicine. This will maximize the value of the medicine and decrease the likelihood of overdose.
3. Improvements in the drug discovery and approval process. The discovery of potential therapies will be made easier using genome targets. Genes have been associated with numerous diseases and disorders. With modern biotechnology, these genes can be used as targets for the development of effective new therapies, which could significantly shorten the drug discovery process.
4. Better vaccines. Safer vaccines can be designed and produced by organisms transformed by means of genetic engineering. These vaccines will elicit the immune response without the attendant risks of infection. They will be inexpensive, stable, easy to store, and capable of being engineered to carry several strains of pathogen at once.

### *Pharmaceutical Products*

Most traditional pharmaceutical drugs are relatively simple molecules that have been found primarily through trial and error to treat the symptoms of a disease or illness. Biopharmaceuticals are large biological molecules known as proteins and these usually target the underlying mechanisms and pathways of a malady (but not always, as is the case with using insulin to treat type 1 diabetes mellitus, as that treatment merely addresses the symptoms of the disease, not the underlying cause which is autoimmunity); it is a relatively young industry. They can deal with targets in humans that may not be accessible with traditional medicines. A patient typically is dosed with a small molecule *via* a tablet while a large molecule is typically injected.

Small molecules are manufactured by chemistry but larger molecules are created by living cells such as those found in the human body: for example, bacteria cells, yeast cells, animal or plant cells. Modern biotechnology is often associated with the use of genetically altered microorganisms such as *E. coli* or yeast for the production of substances like synthetic insulin or antibiotics.

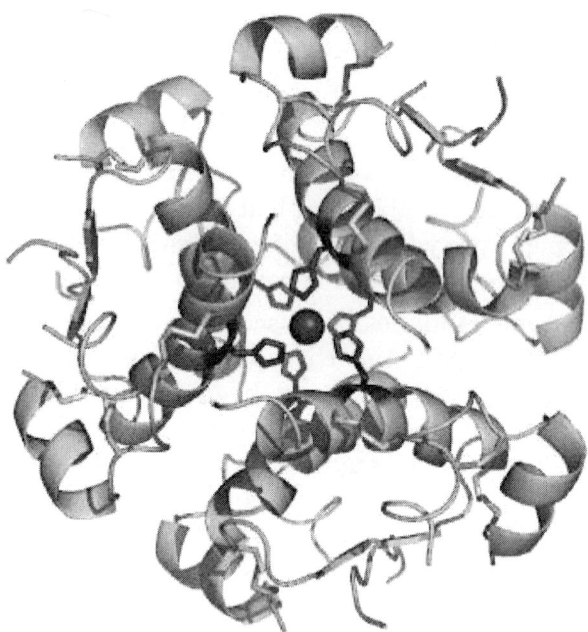

***Figure:*** *Computer-generated image of insulin hexamers highlighting the threefold symmetry, the zinc ions holding it together, and the histidine residues involved in zinc binding.*

It can also refer to transgenic animals or transgenic plants, such as Bt corn. Genetically altered mammalian cells, such as Chinese Hamster Ovary (CHO) cells, are also used to manufacture certain pharmaceuticals. Another promising new biotechnology application is the development of plant-made pharmaceuticals. Biotechnology is also commonly associated with landmark breakthroughs in new medical therapies to treat hepatitis B, hepatitis C, cancers, arthritis, haemophilia, bone fractures, multiple sclerosis, and cardiovascular disorders. The biotechnology industry has also been instrumental in developing molecular diagnostic devices that can be used to define the target patient population for a given biopharmaceutical. Herceptin, for example, was the first drug approved for use with a matching diagnostic test and is used to treat breast cancer in women whose cancer cells express the protein HER2.

Modern biotechnology can be used to manufacture existing medicines relatively easily and cheaply. The first genetically engineered products were medicines designed to treat human diseases. To cite one example, in 1978 Genentech developed synthetic humanized insulin by joining its gene with a plasmid vector inserted into the bacterium *Escherichia coli*. Insulin, widely used for the treatment of diabetes, was previously extracted from the pancreas of abattoir animals (cattle and/or pigs). The resulting genetically engineered bacterium enabled the production of vast quantities of synthetic human insulin at relatively low cost.

According to a 2003 study undertaken by the International Diabetes Federation (IDF) on the access to and availability of insulin in its member countries, synthetic 'human' insulin is considerably more expensive in most countries where both synthetic 'human' and animal insulin are commercially available: e.g. within European countries the average price of synthetic 'human' insulin was twice as high as the price of pork insulin. Yet in its position statement, the IDF writes that "there is no overwhelming evidence to prefer one species of insulin over another" and "[modern, highly purified] animal insulins remain a perfectly acceptable alternative.

Modern biotechnology has evolved, making it possible to produce more easily and relatively cheaply human growth hormone, clotting factors for hemophiliacs, fertility drugs, erythropoietin and other drugs. Most drugs today are based on about 500 molecular targets. Genomic knowledge of the genes involved in diseases, disease pathways, and drug-response sites are expected to lead to the discovery of thousands more new targets.

## Genetic Testing

Genetic testing involves the direct examination of the DNA molecule itself. A scientist scans a patient's DNA sample for mutated sequences.

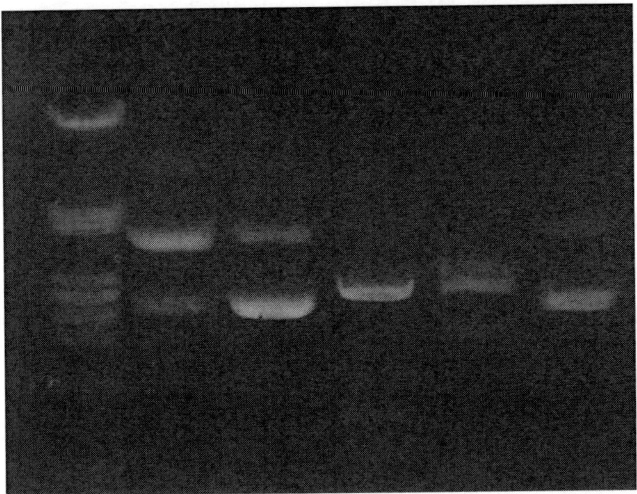

***Figure:*** *Gel electrophoresis*

There are two major types of gene tests. In the first type, a researcher may design short pieces of DNA ("probes") whose sequences are complementary to the mutated sequences. These probes will seek their complement among the base pairs of an individual's genome. If the mutated sequence is present in the patient's genome, the probe will bind to it and flag the mutation. In the second type, a researcher may conduct the gene test by comparing the sequence of DNA bases in a patient's gene to disease in healthy individuals or their progeny.

Genetic testing is now used for:

- Carrier screening, or the identification of unaffected individuals who carry one copy of a gene for a disease that requires two copies for the disease to manifest;
- Confirmational diagnosis of symptomatic individuals;
- Determining sex;
- Forensic/identity testing;
- Newborn screening;
- Prenatal diagnostic screening;
- Presymptomatic testing for estimating the risk of developing adult-onset cancers;
- Presymptomatic testing for predicting adult-onset disorders.

Some genetic tests are already available, although most of them are used in developed countries. The tests currently available can detect mutations associated with rare genetic disorders like cystic fibrosis, sickle cell anemia, and Huntington's disease.

Recently, tests have been developed to detect mutation for a handful of more complex conditions such as breast, ovarian, and colon cancers. However, gene tests may not detect every mutation associated with a particular condition because many are as yet undiscovered, and the ones they do detect may present different risks to different people and populations.

## Controversial Questions

The absence of privacy and anti-discrimination legal protections in most countries can lead to discrimination in employment or insurance or other use of personal genetic information. This raises questions such as whether genetic privacy is different from medical privacy.

1. *Reproductive Issues:* These include the use of genetic information in reproductive decision-making and the possibility of genetically altering reproductive cells that may be passed on

to future generations. For example, germline therapy changes the genetic make-up of an individual's descendants. Thus, any error in technology or judgment may have far-reaching consequences (though the same can also happen through natural reproduction). Ethical issues like designed babies and human cloning have also given rise to controversies between and among scientists and bioethicists, especially in the light of past abuses with eugenics.

2. *Clinical Issues:* These center on the capabilities and limitations of doctors and other health-service providers, people identified with genetic conditions, and the general public in dealing with genetic information.
3. *Effects on Social Institutions:* Genetic tests reveal information about individuals and their families. Thus, test results can affect the dynamics within social institutions, particularly the family.
4. Conceptual and philosophical implications regarding human responsibility, free will vis-à-vis genetic determinism, and the concepts of health and disease.

## Gene Therapy

Gene therapy may be used for treating, or even curing, genetic and acquired diseases like cancer and AIDS by using normal genes to supplement or replace defective genes or to bolster a normal function such as immunity. It can be used to target somatic (i.e., body) or gametes (i.e., egg and sperm) cells. In somatic gene therapy, the genome of the recipient is changed, but this change is not passed along to the next generation. In contrast, in germline gene therapy, the egg and sperm cells of the parents are changed for the purpose of passing on the changes to their offspring.

There are basically two ways of implementing a gene therapy treatment:

1. *Ex vivo*, which means "outside the body" – Cells from the patient's blood or bone marrow are removed and grown in the laboratory. They are then exposed to a virus carrying the desired gene. The virus enters the cells, and the desired gene becomes part of the DNA of the cells. The cells are allowed to grow in the laboratory before being returned to the patient by injection into a vein.
2. *In vivo*, which means "inside the body" – No cells are removed from the patient's body. Instead, vectors are used to deliver the desired gene to cells in the patient's body.

As of June 2001, more than 500 clinical gene-therapy trials involving about 3,500 patients have been identified worldwide. Around 78% of these are in the United States, with Europe having 18%. These trials focus on various types of cancer, although other multigenic diseases are being studied as well. Recently, two children born with severe combined immunodeficiency disorder ("SCID") were reported to have been cured after being given genetically engineered cells.

Gene therapy faces many obstacles before it can become a practical approach for treating disease. At least four of these obstacles are as follows:

1. *Gene delivery tools*. Genes are inserted into the body using gene carriers called vectors. The most common vectors now are viruses, which have evolved a way of encapsulating and delivering their genes to human cells in a pathogenic manner. Scientists manipulate the genome of the virus by removing the disease-causing genes and inserting the therapeutic genes. However, while viruses are effective, they can introduce problems like toxicity, immune and inflammatory responses, and gene control and targeting issues. In addition, in order for gene therapy to provide permanent therapeutic effects, the introduced gene needs to be integrated within the host cell's genome. Some viral vectors effect this in a random fashion, which can introduce other problems such as disruption of an endogenous host gene.
2. *High costs*. Since gene therapy is relatively new and at an experimental stage, it is an expensive treatment to undertake. This explains why current studies are focused on illnesses commonly found in developed countries, where more people can afford to pay for treatment. It may take decades before developing countries can take advantage of this technology.
3. *Limited knowledge of the functions of genes*. Scientists currently know the functions of only a few genes. Hence, gene therapy can address only some genes that cause a particular disease. Worse, it is not known exactly whether genes have more than one function, which creates uncertainty as to whether replacing such genes is indeed desirable.
4. *Multigene disorders and effect of environment*. Most genetic disorders involve more than one gene. Moreover, most diseases involve the interaction of several genes and the environment. For example, many people with cancer not only inherit the disease gene for the disorder, but may have also failed to inherit

specific tumor suppressor genes. Diet, exercise, smoking and other environmental factors may have also contributed to their disease.

## The Next Human Genome Project: Our Microbes

A proposed project to sequence the microorganisms that inhabit our bodies could have a huge impact on human health.

Much as we might like to ignore them, microbes have colonized almost every inch of our bodies, living in our mouths, skin, lungs, and gut. Indeed, the human body has 10 times as many microbial cells as human cells. They're a vital part of our health, breaking down otherwise indigestible foods, making essential vitamins, and even shaping our immune system. Recent research suggests that microbes play a role in diseases, such as ulcers, heart disease, and obesity.

While microbes make up such an intimate part of us, most of our microbial inhabitants remain a mystery. The bacteria in the human body are very difficult to study, since only about 1 percent of them can be grown in the lab. Now a proposed new project to sequence all our microbial residents could change that.

> *"This is completely unexplored territory that is likely to have a large impact on our understanding of human health and disease," says George Weinstock, codirector of the Human Sequencing Center at the Baylor College of Medicine, in Houston. "We hadn't been able to approach it because of the scale of the problem. But now we are finally able to open that door."*

Thanks to ever-improving methods to sequence DNA, scientists can now analyse the genomes of entire microbial communities, a field known as metagenomics. By comparing microbial communities in people of different ages, origins, and health statuses, researchers hope to find out precisely how microorganisms prevent or increase risk for certain diseases and whether they can be manipulated to improve health.

Several metagenomics projects are under way or have been completed, including analysis of the microbes living in the human gut and on the skin. But a true snapshot of our microbial menagerie will require a massive effort, along the lines of the Human Genome Project. "Even though a microbial genome is one-thousandth the size of the human genome, the total number of microbial genes in [the human] body is much greater than human genes because you have so many different species," says Weinstock.

The National Institutes of Health (NIH) is now considering such a project. Metagenomics experts and government officials met last week to determine if the proposal, dubbed the human microbiome, will become an NIH "Roadmap" initiative. These NIH-wide programs identify major gaps in biomedical research and provide financial support on a much larger scale than typical grants. Recent research from Gordon's lab hints at the potential public-health impact of a clearer understanding of our microbial tenants. Gordon and his colleagues have shown that obese people harbor different microbial communities than lean people. And as obese people lost weight, their microbes began to look more like their lean counterparts' microbes.

Researchers aren't yet sure what triggers the differences, but they found in a similar study in mice that the microbial populations of obese mice could more effectively release calories from food during digestion than could microbes of their lean littermates. While exciting, Gordon's research also illustrates the challenges of cataloging microbes. To truly interpret the human microbiome, scientists will need to look at the variation in microbial communities among many people and a variety of populations. Complicating the problem is that, while an individual's human genome is static, a person's microbial composition—and thus his or her microbiome—fluctuates over time. So an accurate picture of one person's microbiome could require multiple resequencing efforts. These types of studies could yield the biggest reward, revealing whether different organisms are correlated with different health states. Gordon and others hope that a microbial analysis will ultimately become a routine part of medical exams, perhaps used to diagnose different diseases. Scientists are still debating whether the microbiome will become a road-map project, and if so, what the final goals of the project will be: should they focus on generating complete sequences of dominant microbes, for example, or devote equal energy to the complex task of studying microbial variation?

## Molecular Genetics in Clinical Practice

### Medical Genetics

Medical genetics is the specialty of medicine that involves the diagnosis and management of hereditary disorders. Medical genetics differs from Human genetics in that human genetics is a field of scientific research that may or may not apply to medicine, but medical genetics refers to the application of genetics to medical care. For example, research on the causes and inheritance of genetic disorders would be considered within both human genetics and medical genetics, while the diagnosis, management, and counseling of individuals with genetic disorders would be considered part of medical genetics.

In contrast, the study of typically non-medical phenotypes such as the genetics of eye colour would be considered part of human genetics, but not necessarily relevant to medical genetics (except in situations such as albinism). *Genetic medicine* is a newer term for medical genetics and incorporates areas such as gene therapy, personalized medicine, and the rapidly emerging new medical specialty, predictive medicine.

## Scope

Medical genetics encompasses many different areas, including clinical practice of physicians, genetic counsellors, and nutritionists, clinical diagnostic laboratory activities, and research into the causes and inheritance of genetic disorders. Examples of conditions that fall within the scope of medical genetics include birth defects and dysmorphology, mental retardation, autism, metabolic and mitochondrial disorders, skeletal dysplasia, connective tissue disorders, cancer genetics, teratogens, and prenatal diagnosis. Medical genetics is increasingly becoming relevant to many common diseases. Overlaps with other medical specialties are beginning to emerge, as recent advances in genetics are revealing etiologies for neurologic, endocrine, cardiovascular, pulmonary, ophthalmologic, renal, psychiatric, and dermatologic conditions.

## Subspecialties

In some ways, many of the individual fields within medical genetics are hybrids between clinical care and research. This is due in part to recent advances in science and technology that have enabled an unprecedented understanding of genetic disorders.

## Clinical Genetics

Clinical genetics is the practice of clinical medicine with particular attention to hereditary disorders. Referrals are made to genetics clinics for a variety of reasons, including birth defects, developmental delay, autism, epilepsy, short stature, and many others. Examples of genetic syndromes that are commonly seen in the genetics clinic include chromosomal rearrangements, Down syndrome, DiGeorge syndrome (22q11.2 Deletion Syndrome), Fragile X syndrome, Marfan syndrome, Neurofibromatosis, Turner syndrome, and Williams syndrome.

## Metabolic/Biochemical Genetics

Metabolic (or biochemical) genetics involves the diagnosis and management of inborn errors of metabolism in which patients have enzymatic deficiencies that perturb biochemical pathways involved in metabolism of carbohydrates, amino acids, and lipids. Examples of metabolic disorders include galactosemia, glycogen storage disease, lysosomal storage disorders, metabolic acidosis, peroxisomal disorders, phenylketonuria, and urea cycle disorders.

### Cytogenetics

Cytogenetics is the study of chromosomes and chromosome abnormalities. While cytogenetics historically relied on microscopy to analyse chromosomes, new molecular technologies such as array comparative genomic hybridization are now becoming widely used. Examples of chromosome abnormalities include aneuploidy, chromosomal rearrangements, and genomic deletion/duplication disorders.

### Molecular Genetics

Molecular genetics involves the discovery of and laboratory testing for DNA mutations that underlie many single gene disorders. Examples of single gene disorders include achondroplasia, cystic fibrosis, Duchenne muscular dystrophy, hereditary breast cancer (BRCA1/2), Huntington disease, Marfan syndrome, Noonan syndrome, and Rett syndrome. Molecular tests are also used in the diagnosis of syndromes involving epigenetic abnormalities, such as Angelman syndrome, Beckwith-Wiedemann syndrome, Prader-willi syndrome, and uniparental disomy.

### Mitochondrial Genetics

Mitochondrial genetics concerns the diagnosis and management of mitochondrial disorders, which have a molecular basis but often result in biochemical abnormalities due to deficient energy production. There exists some overlap between medical genetic diagnostic laboratories and molecular pathology.

### Genetic Counseling

Genetic counseling is the process of providing information about genetic conditions, diagnostic testing, and risks in other family members, within the framework of nondirective counseling. Genetic counsellors are non-physician members of the medical genetics team who specialize in family risk assessment and counseling of patients regarding genetic disorders. The precise role of the genetic counselor varies somewhat depending on the disorder.

### History

Although genetics has its roots back in the 19th century with the work of the Bohemian monk Gregor Mendel and other pioneering scientists, human genetics emerged later. It started to develop, albeit slowly, during the first half of the 20th century. Mendelian (single-gene) inheritance was studied in a number of important disorders such as albinism, brachydactyly (short fingers and toes), and hemophilia. Mathematical approaches were also devised and applied to human genetics. Population genetics was created. Medical genetics was a late developer, emerging largely after the close of World War II (1945) when the eugenics movement had fallen into disrepute. The Nazi misuse of eugenics sounded its death knell. Shorn of eugenics, a scientific approach could be used and was applied to human and medical genetics. Medical genetics saw an increasingly rapid rise in the second half of the 20th century and continues in the 21st century.

### Current Practice

The clinical setting in which patients are evaluated determines the scope of practice, diagnostic, and therapeutic interventions. For the purposes of general discussion, the typical encounters between patients and genetic practitioners may involve:

- Referral to an out-patient genetics clinic (pediatric, adult, or combined) or an in-hospital consultation, most often for diagnostic evaluation.
- Specialty genetics clinics focusing on management of inborn errors of metabolism, skeletal dysplasia, or lysosomal storage diseases.
- Referral for counseling in a prenatal genetics clinic to discuss risks to the pregnancy (advanced maternal age, teratogen exposure, family history of a genetic disease), test results (abnormal maternal serum screen, abnormal ultrasound), and/or options for prenatal diagnosis (typically amniocentesis or chorionic villus sampling).
- Multidisciplinary specialty clinics that include a clinical geneticist or genetic counselor (cancer genetics, cardiovascular genetics, craniofacial or cleft lip/palate, hearing loss clinics, muscular dystrophy/neurodegenerative disorder clinics).

### Diagnostic Evaluation

Each patient will undergo a diagnostic evaluation tailored to their own particular presenting signs and symptoms. The geneticist will establish a differential diagnosis and recommend appropriate testing. Increasingly, clinicians use SimulConsult, paired with the National Library of Medicine Gene Review articles, to narrow the list of hypotheses (known as the differential diagnosis) and identify the tests that are relevant for a particular patient. These tests might evaluate for chromosomal disorders, inborn errors of metabolism, or single gene disorders.

### Chromosome Studies

Chromosome studies are used in the general genetics clinic to determine a cause for developmental delay/mental retardation, birth defects, dysmorphic features, and/or autism. Chromosome analysis is also performed in the prenatal setting to determine whether a fetus is affected with aneuploidy or other chromosome rearrangements. Finally, chromosome abnormalities are often detected in cancer samples. A large number of different methods have been developed for chromosome analysis:

- Chromosome analysis using a karyotype involves special stains that generate light and dark bands, allowing identification of each chromosome under a microscope.
- Fluorescence in situ hybridization (FISH) involves fluorescent labelling of probes that bind to specific DNA sequences, used for identifying aneuploidy, genomic deletions or duplications, characterizing chromosomal translocations and determining the origin of ring chromosomes.
- Chromosome painting is a technique that uses fluorescent probes specific for each chromosome to differentially label each chromosome. This technique is more often used in cancer cytogenetics, where complex chromosome rearrangements can occur.
- Array comparative genomic hybridization is a new molecular technique that involves hybridization of an individual DNA sample to a glass slide or microarray chip containing molecular probes (ranging from large ~200kb bacterial artificial chromosomes to small oligonucleotides) that represent unique regions of the genome. This method is particularly sensitive for detection of genomic gains or losses across the genome but does not detect balanced translocations or distinguish the location of duplicated genetic material (for example, a tandem duplication versus an insertional duplication).

### Basic Metabolic Studies

Biochemical studies are performed to screen for imbalances of metabolites in the bodily fluid, usually the blood (plasma/serum) or urine, but also in cerebrospinal fluid (CSF). Specific tests of enzyme function (either in leukocytes, skin fibroblasts, liver, or muscle) are also employed under certain circumstances. In the US, the newborn screen incorporates biochemical tests to screen for treatable conditions such as galactosemia and phenylketonuria (PKU). Patients suspected to have a metabolic condition might undergo the following tests:

- Quantitative amino acid analysis is typically performed using the ninhydrin reaction, followed by liquid chromatography to measure the amount of amino acid in the sample (either urine, plasma/serum, or CSF). Measurement of amino acids in plasma or serum is used in the evaluation of disorders of amino acid metabolism such as urea cycle disorders, maple syrup urine disease, and PKU. Measurement of amino acids in urine can be useful in the diagnosis of cystinuria or renal Fanconi syndrome as can be seen in cystinosis.
- Urine organic acid analysis can be either performed using quantitative or qualitative methods, but in either case the test is used to detect the excretion of abnormal organic acids. These compounds are normally produced during bodily metabolism of amino acids and odd-chain fatty acids, but accumulate in patients with certain metabolic conditions.
- The acylcarnitine combination profile detects compounds such as organic acids and fatty acids conjugated to carnitine. The test is used for detection of disorders involving fatty acid metabolism, including MCAD.
- Pyruvate and lactate are byproducts of normal metabolism, particularly during anaerobic metabolism. These compounds normally accumulate during exercise or ischemia, but are also elevated in patients with disorders of pyruvate metabolism or mitochondrial disorders.
- Ammonia is an end product of amino acid metabolism and is converted in the liver to urea through a series of enzymatic reactions termed the urea cycle. Elevated ammonia can therefore be detected in patients with urea cycle disorders, as well as other conditions involving liver failure.
- Enzyme testing is performed for a wide range of metabolic disorders to confirm a diagnosis suspected based on screening tests.

### *Molecular Studies*

- DNA sequencing is used to directly analyse the genomic DNA sequence of a particular gene. In general, only the parts of the gene that code for the expressed protein (exons) and small amounts of the flanking untranslated regions and introns are analysed. Therefore, although these tests are highly specific and sensitive, they do not routinely identify all of the mutations that could cause disease.

- DNA methylation analysis is used to diagnose certain genetic disorders that are caused by disruptions of epigenetic mechanisms such as genomic imprinting and uniparental disomy.
- Southern blotting is an early technique basic on detection of fragments of DNA separated by size through gel electrophoresis and detected using radio labeled probes. This test was routinely used to detect deletions or duplications in conditions such as Duchenne muscular dystrophy but is being replaced by high-resolution array comparative genomic hybridization techniques. Southern blotting is still useful in the diagnosis of disorders caused by trinucleotide repeats.
- Short tandem repeats are unique markers that can be used to determine haplotypes and are used in identity testing for maternal cell contamination.

### *Treatments*

Each cell of the body contains the hereditary information (DNA) wrapped up in structures called chromosomes. Since genetic syndromes are typically the result of alterations of the chromosomes or genes, there is no treatment currently available that can correct the genetic alterations in every cell of the body.

Therefore, there is currently no "cure" for genetic disorders. However, for many genetic syndromes there is treatment available to manage the symptoms. In some cases, particularly inborn errors of metabolism, the mechanism of disease is well understood and offers the potential for dietary and medical management to prevent or reduce the long-term complications. In other cases, infusion therapy is used to replace the missing enzyme. Current research is actively seeking to use gene therapy or other new medications to treat specific genetic disorders.

### *Career Paths and Training*

There are a variety of career paths within the field of medical genetics, and naturally the training required for each area differs considerably. It should be noted that the information included in this section applies to the typical pathways in the United States and there may be differences in other countries. US Practitioners in clinical, counseling, or diagnostic subspecialties generally obtain board certification through the American Board of Medical Genetics.

### *Ethical, Legal and Social Implications*

Genetic information provides a unique type of knowledge about an individual and his/her family, fundamentally different than a typically laboratory test that provides a "snapshot" of an individual's health status. The unique status of genetic information and inherited disease has a number of ramifications with regard to ethical, legal, and societal concerns.

### Societies

The more empirical approach to human and medical genetics was formalized by the founding in 1948 of the American Society of Human Genetics. The Society first began annual meetings that year (1948) and its international counterpart, the International Congress of Human Genetics, has met every 5 years since its inception in 1956. The Society publishes the American Journal of Human Genetics on a monthly basis. Medical genetics is now recognized as a distinct medical specialty in the U.S. with its own approved board (the American Board of Medical Genetics) and clinical specialty college (the American College of Medical Genetics). The College holds an annual scientific meeting, publishes a monthly journal, Genetics in Medicine, and issues position papers and clinical practice guidelines on a variety of topics relevant to human genetics.

### Research

The broad range of research in medical genetics reflects the overall scope of this field, including basic research on genetic inheritance and the human genome, mechanisms of genetic and metabolic disorders, translational research on new treatment modalities, and the impact of genetic testing.

### Basic Genetics Research

Basic research geneticists usually undertake research in universities, biotechnology firms and research institutes.

### Allelic Architecture of Disease

Sometimes the link between a disease and an unusual gene variant is more subtle. The genetic architecture of common diseases is an important factor in determining the extent to which patterns of genetic variation influence group differences in health outcomes. According to the common disease/common variant hypothesis, common variants present in the ancestral population before the dispersal of modern humans from Africa play an important role in human diseases. Genetic variants associated with Alzheimer disease, deep venous thrombosis, Crohn disease, and type 2 diabetes appear to adhere to this model. However, the generality of the model has not yet been established and, in some cases, is in doubt. Some diseases, such as many common cancers, appear not to be well described by the common disease/common variant model.

Another possibility is that common diseases arise in part through the action of combinations of variants that are individually rare. Most of the disease-associated alleles discovered to date have been rare, and rare variants are more likely than common variants to be differentially distributed among groups distinguished by ancestry. However, groups could harbor different, though perhaps overlapping, sets of rare variants, which would reduce contrasts between groups in the incidence of the disease.

The number of variants contributing to a disease and the interactions among those variants also could influence the distribution of diseases among groups. The difficulty that has been encountered in finding contributory alleles for complex diseases and in replicating positive associations suggests that many complex diseases involve numerous variants rather than a moderate number of alleles, and the influence of any given variant may depend in critical ways on the genetic and environmental background. If many alleles are required to increase susceptibility to a disease, the odds are low that the necessary combination of alleles would become concentrated in a particular group purely through drift.

### Population Substructure in Genetics Research

One area in which population categories can be important considerations in genetics research is in controlling for confounding between population substructure, environmental exposures, and health outcomes. Association studies can produce spurious results if cases and controls have differing allele frequencies for genes that are not related to the disease being studied, although the magnitude of this problem in genetic association studies is subject to debate. Various methods have been developed to detect and account for population substructure, but these methods can be difficult to apply in practice.

Population substructure also can be used to advantage in genetic association studies. For example, populations that represent recent mixtures of geographically separated ancestral groups can exhibit longer-range linkage disequilibrium between susceptibility alleles and genetic markers than is the case for other populations. Genetic studies can use this admixture linkage disequilibrium to search for disease alleles with fewer markers than would be needed otherwise.

Association studies also can take advantage of the contrasting experiences of racial or ethnic groups, including migrant groups, to search for interactions between particular alleles and environmental factors that might influence health.

## Inborn Error of Metabolism

Inborn errors of metabolism comprise a large class of genetic diseases involving disorders of metabolism. The majority are due to defects of single genes that code for enzymes that facilitate conversion of various substances (substrates) into others (products). In most of the disorders, problems arise due to accumulation of substances which are toxic or interfere with normal function, or to the effects of reduced ability to synthesize essential compounds. Inborn errors of metabolism are now often referred to as congenital metabolic diseases or inherited metabolic diseases.

The term *inborn error of metabolism* was coined by a British physician, Archibald Garrod (1857–1936), in the early 20th century (1908). He is known for work that prefigured the "one gene-one enzyme" hypothesis, based on his studies on the nature and inheritance of alkaptonuria. His seminal text, *Inborn Errors of Metabolism* was published in 1923.

### *Major Categories of Inherited Metabolic Diseases*

Traditionally the inherited metabolic diseases were categorized as disorders of carbohydrate metabolism, amino acid metabolism, organic acid metabolism, or lysosomal storage diseases. In recent decades, hundreds of new inherited disorders of metabolism have been discovered and the categories have proliferated. Following are some of the major classes of congenital metabolic diseases, with prominent examples of each class. Many others do not fall into these categories. ICD-10 codes are provided where available.

- Disorders of carbohydrate metabolism
    - E.g., glycogen storage disease
- Disorders of amino acid metabolism
    - E.g., phenylketonuria, maple syrup urine disease, glutaric acidemia type 1
- Disorders of organic acid metabolism (organic acidurias)
    - E.g., alcaptonuria
- Disorders of fatty acid oxidation and mitochondrial metabolism
    - E.g., medium chain acyl dehydrogenase deficiency (glutaric acidemia type 2)
- Disorders of porphyrin metabolism
    - E.g., acute intermittent porphyria

- Disorders of purine or pyrimidine metabolism
  - E.g., Lesch-Nyhan syndrome
- Disorders of steroid metabolism
  - E.g., congenital adrenal hyperplasia
- Disorders of mitochondrial function
  - E.g., Kearns-Sayre syndrome
- Disorders of peroxisomal function
  - E.g., Zellweger syndrome
- Lysosomal storage disorders
  - E.g., Gaucher's disease
  - E.g., Niemann Pick disease.

### *Incidence*

In a study in British Columbia, the overall incidence of the inborn errors of metabolism were estimated to be 70 per 100,000 live births or 1 in 1,400 births, overall representing more than approximately 15% of single gene disorders in the population.

### *Manifestations and Presentations*

Because of the enormous number of these diseases and wide range of systems affected, nearly every "presenting complaint" to a doctor may have a congenital metabolic disease as a possible cause, especially in childhood. The following are examples of potential manifestations affecting each of the major organ systems:

- Growth failure, failure to thrive, weight loss
- Ambiguous genitalia, delayed puberty, precocious puberty
- Developmental delay, seizures, dementia, encephalopathy, stroke
- Deafness, blindness, pain agnosia
- Skin rash, abnormal pigmentation, lack of pigmentation, excessive hair growth, lumps and bumps
- Dental abnormalities
- Immunodeficiency, thrombocytopenia, anemia, enlarged spleen, enlarged lymph nodes
- Many forms of cancer
- Recurrent vomiting, diarrhea, abdominal pain
- Excessive urination, renal failure, dehydration, edema
- Hypotension, heart failure, enlarged heart, hypertension, myocardial infarction

- Hepatomegaly, jaundice, liver failure
- Unusual facial features, congenital malformations
- Excessive breathing (hyperventilation), respiratory failure
- Abnormal behaviour, depression, psychosis
- Joint pain, muscle weakness, cramps
- Hypothyroidism, adrenal insufficiency, hypogonadism, diabetes mellitus.

## Diagnostic Techniques

Dozens of congenital metabolic diseases are now detectable by newborn screening tests, especially the expanded testing using mass spectrometry. This is an increasingly common way for the diagnosis to be made and sometimes results in earlier treatment and a better outcome. There is a revolutionary GC/MS based technology with an integrated analytics system, which has now made it possible to test a newborn for over 100 genetic metabolic disorders.

Because of the multiplicity of conditions, many different diagnostic tests are used for screening. An abnormal result is often followed by a subsequent "definitive test" to confirm the suspected diagnosis.

# Chapter 5

# Impact of Genomics and Bioinformatics on Biotechnology

The major societal problems include labour intensive agriculture, crop losses due to various factors, malnutrition, prevalence of infectious diseases due to pathogens, and diseases due to biochemical and physiological disorders. When coupled with the teeming population and persistent degradation of the environment, these problems become truly magnified to an enormous complexity.

The discovery of genes as the basic units of inheritance and DNA as the macromolecule carrying genes in all living organisms, with just a few exceptions, has given a tremendous boost for biotechnology to grow and aid the society, in its modern form (mediated by DNA, RNA and Protein) as a highly research intensive and application sector. About five to ten years ago, the focus of agricultural and biomedical research had been generally on a single isolated gene/enzyme/protein/event and many of these efforts led to successful discovery of diagnostic and corrective measures against a few of the challenges mentioned above. However, in recent times, owing to the continuous spectacular developments in the areas of macromolecular chemistry, macromolecule sequencing machines and computational tools to analyse biological data, the focus has shifted from single to multiple gene/protein/enzyme/events to assist rapid discovery of remedies for the prevailing problems.

The genome and the deduced protein sequence for a large number of organisms are now available for functional analysis and confirmation at the biochemical pathway or physiological process levels, which enables understanding of the mechanisms of trait development in crops or the occurrence of diseases in many living systems. The DNA information provided is in long strings of four-lettered sequences organized into distinct open reading frames (ORFs) associated with additional sequences for governing these ORFs. The ORFs representing all the genes constitute the genetic make-up of that organism. The ORFs translate into twenty-lettered amino acid sequences, which forms the basic framework to determine the functional properties. While many of the translated sequences from ORFs have known functions assigned to them, there are many that are unknown, awaiting exciting discoveries which in turn may lead to identification of new biological functions of economical and clinical significance. A simultaneous genome-wide analysis of gene expression ("Transcription Profiling" or "Transcriptomics") has been possible by the development of "DNA chip" or the "Micro array" technology. Likewise, a simultaneous genome-wide analysis of the pattern of metabolite synthesis ("Metabolic Profiling" or "Metabolomics") in a living system provides an insight towards understanding the functioning of biochemical pathways at a given point in time. These technologies are high throughput methods developed for understanding at the organism level, the behavioural pattern ("Phenotypic Expression" or "Phenomics") of microbes and higher eukaryotes (including plants and animals) under various environmental conditions.

To help analyse the enormous amount of biological data that gets generated from sequencing projects, transcription and metabolic profiling experiments, Bioinformatics provides software tools for construction of complete catalogues of the ORFs, gene products, functional properties, interaction amongst the gene products, elucidation of pathways, comparison of genomes, similarity searches, and insights towards genetic modification of economically important biological systems to result in new products and new therapies of economical and clinical significance respectively.

Many beneficial as well as harmful traits involve multiple genes, proteins, biochemical pathways and physiological processes. With this in view, the inference of results generated from high throughput analysis at the whole genome level using the tools of bioinformatics would certainly have a tremendous impact on research efforts towards finding solutions for the societal challenges, from the perspective of biotechnology.

## Drug Design Using Molelcular Operating Environemnt

The explosion of information has propelled the rapid development of bioinformatics. The future of postgenomic biology requires something that does not even exist today. There are numerous algorithms and new methodologies that need to be developed consistently. Bioinformatics is reckoned to revolutionize the disease and therapies research.

One has to consistently looking at accelerating the drug discovery process for genomics, proteomics and *in silico* biology. Simulating the *in vivo* to *in silico* is a challenging task today. But today it is possible to achieve with the advancement in High performance computing power and with the software developed my efficient algorithms. There are a few software available for *in silico* drug design. One such product is MOE, Molecular Operating Environment developed by chemical computing group Canada. Which contains all the programs from sequence analysis protein modelling and structure based drug design.

During this talk I would like to give a demo on how this software is helpful in the following areas.

### *Molecular Modelling and Simulations*

- Molecular Builders. Build small molecules, carbohydrates, proteins, and DNA and crystal structures in 3D with a collection of graphical interfaces.
- Molecular Mechanics and Dynamics. Perform large-scale energy minimizations using AMBER '89/'94, MMFF94, PEFSAC95 or Engh-Huber force field parameters with an implicit solvent model.
- Conformational analyses include both stochastic and systematic searches (including rings). Calculate dynamics trajectories in the NVE, NVT, NPT and NPH ensembles.
- Implicit Solvent Electrostatics. Solve the non-linear Poisson-Boltzmann equation with a non-linear multi-grid method to produce the implicit solvent electrostatics field.
- Molecule Alignment. Align a collection (flexibly or rigid-body) of small molecules that are presumed to have similar biological activity. Alignments are useful for the elucidation of pharmacophores, input to comparative field analysis and template forcing.

### *High Throughput Drug Discovery*

- QSAR. Create large-scale linear or probabilistic predictive models of biological activity (or other properties). Use CCG's patent-pending Binary-QSAR method for HTS data analysis and ADME-based classification.
- Combinatorial Library Design. Design combinatorial libraries using either maximum diversity or QSAR-biased methodologies.

- Molecular Descriptors. Calculate over 480 molecular properties including topological indices, octanol/water logP, molar refractivity and CCG's VSA descriptors that have shown wide applicability in compound classification, QSAR and ADME property modelling.
- Compound Clustering. A powerful non-parametric subdivision of probability space technique is used to rapidly cluster large databases. Compounds can be clustered by fingerprints (3D Pharmacophore, graph pharmacophore or MACCS keys) using the Jarvis-Patrick method.
- Diverse Subset. Calculate maximal diverse subsets of compounds based on 3D conformations, molecular descriptors or molecular fingerprints

## *Protein Modelling and Bioinformatics*

- Protein Structure Database. CCG has created a searchable database of over 15,000 protein structures from the Protein Data Bank. Each structure was imported using CCG's PDB reader, that corrects many of the errors commonly found in protein structure files.
- Fold Identification. CCG has created a searchable database of structural families of proteins by exhaustively and iteratively clustering the Protein Data Bank. Incorporating rigorous sequence-to-family alignment, secondary structure prediction, hydrophobic fitness and Z-scoring, the search procedure
- Multiple Alignments. CCG's unique technology for simultaneous multiple sequence and structure alignment superposition with no "master" sequence.
- Consensus Features. Geometric criteria are used to rapidly determine the structurally conserved features in a family of proteins including conserved waters.
- Structure Prediction. Homology modelling techniques are used to build complete high-quality 3D structures of proteins from a template.
- Protein Mechanics and Dynamics. Proteins structures can be refined using either AMBER '89, '94, MMFF94 or Engh-Huber parameters augmented with an implicit solvent model. Molecular dynamics can be performed in either the NVE, NVT, NPT or NPH ensembles.

- Stereo chemical Quality. Statistical measures of bond lengths, angles, backbone dihedrals and non-bonded contacts are used to assess the overall stereo chemical quality of a protein structure.
- Contact Analysis. Stabilizing contacts such as hydrogen bonds, salt bridges, hydrophobic contacts and disulfide bonds are often implicated in protein function.

### *Structure Based Drug Design*

- *Structure Preparation:* Automatically connect and assign type information using element and coordinate information. Add hydrogens and refine/relax structures using AMBER '89, '94, MMFF94 or Engh-Huber parameters augmented with an implicit solvent model.
- *Visualization:* Receptors and ligands can be displayed in a variety of styles including line, stick and CPK with full control over colours. Connolly, Accessible and van der Waals interaction surfaces can be coloured in a number of ways including by Pocket and Hydrogen Bond. Both split-pair and full stereo modes are supported.
- *Active Site Detection:* An alpha shape algorithm is used to determine potential active sites in 3D protein structures.
- *Ligand Docking:* A simulated annealing search algorithm is used to flexibly dock a ligand into the active site of a receptor in an effort to predict the binding conformation. A grid-based energy evaluation is used to score the docked conformations.
- *MultiFragment:* Search helps to understand the interactions between chemical functional groups with the active site of a receptor. An ensemble of fragments are randomly placed into the active site and subjected to a special energy minimization protocol to determine the preferred locations of each functional group.

## Human Genome Data in Relation to Disease Management and Control

Cells are the fundamental working units of every living system. All the instructions needed to direct their activities are contained within deoxyribonucleic acid (DNA). A genome is all the DNA in an organism, including its genes. Genes carry information for making proteins required by all organisms. It is the proteins that perform most life functions and even make up the majority of cellular structures. DNA is made up of four bases A, T, C & G that are repeated millions of billions of times throughout a genome. The human genome has 3 billion pairs of bases. The particular order of As, Ts, Cs and Gs is extremely important. The order underlies all of life's diversity, even dictating whether an organism is human or another species such as yeast, rice or fruit fly. The human genome project (HGP) is one of the great feats of exploration in history. HGP is coordinated by the U.S. Department of Energy and the National Institutes of Health that officially began in 1990.

The goals of the HGP are to:

- Identify all the approximate 30,000 genes in human DNA.
- Determine the sequences of the 3 billion chemical base pairs that make up human DNA.
- Store this data in databases.
- Improve tools for data analysis.
- Transfer related technologies to the private sector, &
- Address the ethical, legal and social issues (ELS1) that may arise from the project.
- To sequence genomes of model organisms (E.coli, fruit fly and mouse) to help interpret human DNA.

In February 2001, HGP and Celera Genomics scientists published the long-awaited details of the working-draft DNA sequence. An international research effort to sequence and map all of the genes together known as the genome-of the members of our species. Homosapiens will be finished to high quality by 2003. Once complete, it will give us the chance for the first time to read natures complete genetic blueprint for building a human being. The following methodologies are used to sequence a genome;

- Mapping
- Building libraries
- Sub-cloning
- Storing and copying DNA in E.coli
- Preparing DNA for sequencing
- Sequencing reaction and product production
- Separating the sequencing reaction
- Reading the sequencing reaction
- Assembling the results
- Working draft sequence/conclusion.

The working draft tells us the following:

- The human genome contains 3164.7 million chemical nucleotide bases (A, C, T, and G).

- The average gene consists of 3000 bases, but sizes vary greatly, with the largest known human gene being dystrophin at 2.4 million bases.
- The total number of genes is estimated at 30,000 to 35,000 much lower than previous estimates of 80,000 to 140,000 that had been based on extrapolations from gene-rich areas as opposed to a composite of gene-rich and gene-poor areas.
- Almost all (99.9%) nucleotide bases are exactly the same in all people.
- The functions are unknown for over 50% of discovered genes.
- Less than 2% of the genome codes for proteins.
- Repeated sequences that do not code for proteins ("junk DNA") make up at least 50% of the human genome.
- Repetitive sequences are thought to have no direct functions, but they shed light on chromosome structure and dynamics. Over time, these repeats reshape the genome by rearranging it, creating entirely new genes, and modifying and reshuffling existing genes.
- During the past 50 million years, a dramatic decrease seems to have occurred in the rate of accumulation of repeats in the human genome.
- The human genome's gene-dense "urban centres" are predominantly composed of the DNA building blocks G and C.
- In contrast, the gene-poor "deserts" are rich in the DNA building blocks A and T. GC- and AT-rich regions usually can be seen through a microscope as light and dark bands on chromosomes.
- Genes appear to be concentrated in random areas along the genome, with vast expanses of noncoding DNA between.
- Stretches of up to 30,000 C and G bases repeating over and over often occur adjacent to gene-rich areas, forming a barrier between the genes and the "junk DNA." These CpG islands are believed to help regulate gene activity.
- Chromosome 1 has the most genes (2968), and the Y chromosome has the fewest (231).

How the Human Compares with Other Organisms ;

- Unlike the human's seemingly random distribution of gene-rich areas, many other organisms' genomes are more uniform, with genes evenly spaced throughout.

- Humans have on average three times as many kinds of proteins as the fly or worm because of mRNA transcript "alternative splicing" and chemical modifications to the proteins. This process can yield different protein products from the same gene.
- Humans share most of the same protein families with worms, flies, and plants, but the number of gene family members has expanded in humans, especially in proteins involved in development and immunity.
- The human genome has a much greater portion (50%) of repeat sequences than the mustard weed (11%), the worm (7%), and the fly (3%).
- Although humans appear to have stopped accumulating repeated DNA over 50 million years ago, there seems to be no such decline in rodents. This may account for some of the fundamental differences between hominids and rodents, although gene estimates are similar in these species.

Scientists have proposed many theories to explain evolutionary contrasts between humans and other organisms, including those of life span, litter sizes, inbreeding, and genetic drift.

Variations and Mutations ;

- Scientists have identified about 1.4 million locations where single-base DNA differences (SNPs) occur in humans. This information promises to revolutionize the processes of finding chromosomal locations for disease-associated sequences and tracing human history.
- The ratio of germline (sperm or egg cell) mutations is 2:1 in males vs females. Researchers point to several reasons for the higher mutation rate in the male germline, including the greater number of cell divisions required for sperm formation than for eggs.

*Will understanding the human genome transform preventive, diagnostic and therapeutic medicine?*

## Diagnosing and Predicting Disease and Disease Susceptibility

All diseases have a genetic component, whether inherited or resulting from the body's response to environmental stresses like viruses or toxins. The successes of the Human Genome Project (HGP) have even enabled researchers to pinpoint errors in genes—the smallest units of heredity—that cause or contribute to disease.

The ultimate goal is to use this information to develop new ways to treat, cure, or even prevent the thousands of diseases that afflict humankind. But the road from gene identification to effective treatments is long and fraught with challenges. In the meantime, biotechnology companies are racing ahead with commercialization by designing diagnostic tests to detect errant genes in people suspected of having particular diseases or at risk for developing them.

An increasing number of gene tests are becoming available commercially although the scientific community continues to debate the best way to deliver them to the public and medical communities that are often unaware of their scientific and social implications. While some of these tests have greatly improved and even saved lives, scientists remain unsure of how to interpret many of them. Also, patients taking the tests face significant risks of jeopardizing their employment or insurance status. And because genetic information is shared, these risks can extend beyond them to their family members as well.

### *Disease Intervention*

Within the next decade, researchers will find most human genes. Explorations into the function of each one —a major challenge extending far into the 21st century —will shed light on how faulty genes play a role in disease causation. With this knowledge, commercial efforts will shift away from diagnostics and toward developing a new generation of therapeutics based on genes.

Drug design will be revolutionized as researchers create new classes of medicines based on a reasoned approach using gene sequence and protein structure function information rather than the traditional trial-and-error method. The drugs, targeted to specific sites in the body, promise to have fewer side effects than many of today's medicines.

The potential for using genes themselves to treat disease—known as gene therapy—is the most exciting application of DNA science. It has captured the imaginations of the public and the biomedical community for good reason. This rapidly developing field holds great potential for treating or even curing genetic and acquired diseases, using normal genes to replace or supplement a defective gene or to bolster immunity to disease (e.g., by adding a gene that suppresses tumour growth).

One of the greatest impacts of having the sequence may well be in enabling an entirely new approach to biological research. Within 20 years, novel drugs will be available that derive from a detailed molecular understanding of common illnesses like diabetes and high blood pressure. The drugs will be designer therapies that target molecules logically and are therefore potent without significant side effects. Drugs like those for cancer will routinely be matched to a patient's likely response, as predicted by molecular fingerprinting. Diagnoses of many conditions will be much more thorough and specific than now. For example, a patient who learns that he has high cholesterol will also know which genes are responsible, what effect the high cholesterol is likely to have, and what diet and pharmacologic measures will work best for him.

By 2050, many potential diseases will be cured at the molecular level before they arise, though large inequities worldwide in access to these advances will continue to stir tensions. When people become sick, gene therapies and drug therapies will home in on individual genes, as they exist in individual people, making for precise, customized medical treatment. The average life span will reach 90 to 95 years, and a detailed understanding of human aging genes will spur efforts to expand the maximum span of human life.

## Immunology and Immunotechnology

In ancient China and India, there was an effective, though highly dangerous, practice of introducing the fluid from the pustules of small pox (*variola, variolae)* patientsinto healthy individuals, through dermal incisions. This practice, called variolation, using live smallpox virus, was aimed at protecting the individual from contracting the disease.

In 1796, Edward Jenner, the English Physician, obtained the pus from the pustules of a dairymaid suffering from cowpox and introduced it, through a nick made in the arm, into the system of an eight-year-old boy and demonstrated that it gave the boy immunity against smallpox. This has opened up a new area in medicine, immunology.

Immunology is sometimes called serology, as the principal participants of immunological reactions reside in the blood serum, although, strictly speaking, serology is the study of the serum, and the properties and functions of its components.

Rooted in the Latin word *vacca* for cow, the introduced substance is called the vaccine and the process vaccination. Vaccination imparts immunity (protection) against the disease, and so, the individual is immunised. Under the current practices, a killed or attenuated (made less potent) pathogenic organism (viruses, bacteria) or its biological product such as a protein, such as the cholera or tetanus toxin (the inoculum) is introduced into a living being. This process is inoculation. In practice, the terms vaccine and inoculum, are virtually synonymous and so are vaccination and inoculation.

Over the time, immunology has absorbed to an immense benefit, from advances in human physiology, medicine, biochemistry, molecular biology (particularly protein structure and chemistry), and a range of electronic instrumentation including computers. Today, immunology is a very complex and sophisticated area of biology, which has become one of the most versatile research tools in biology and medicine, as well as a powerful weapon in the armoury of prevention and management of several viral and bacterial diseases. The significance of immunology in human is amply reflected by the large number of Nobel Prizes awarded to research in this area. These award winning discoveries represent the mile stones in the development of modern immunology, and constitute more than 10 per cent of all Nobel awards for Physiology or Medicine, since 1901.

Immunology helped us to eradicate smallpox. It has saved millions from polio, cholera, hepatitis, tetanus, rabies, and several other potentially debilitating or fatal diseases. It is the hope of mankind, to prevent many other diseases such as malaria, tuberculosis, and even certain types of cancer, through appropriate vaccines.

Immunotechnology, is an important arm of biotechnology, constituting the industrial scale application of immunological procedures to produce vaccines, for mass immunisation to prevent prevalent diseases and/or producing immunological therapeutic agents to cure the afflicted. Production of protein vaccines has been in large-scale use for a long time and the current trend is to develop the more specific DNA vaccines.

### *Immunological Defensive Response*

When an infecting organism gains entry into the mammalian system for the first time, the immune system of the mammal reacts, mainly in response to the proteins of the invading organism, generally called antigens, by producing a special class of proteins called antibodies. This first encounter is very important because the system gains a 'memory' template of the three dimensional structural configuration of the epitopes of the antigens. This memory may last for a few hours or days (common colds), a few months (cholera, tetanus) or the lifetime (smallpox) of the individual. Antigens, the molecules that trigger antibody response, fall into three categories, called antigens, immunogens or haptens, which are sourced in the protein coats of viruses, cell walls of bacteria, secretions of pathogenic organisms, proteins from plant or animal parts that were injected, ingested or inhaled, or introduced into the system by some means.

When the same antigen gains entry a second and the subsequent times, the immune system recognises the foreign entity and produces antibodies specific to the antigen, basing on the memory, developed on the first encounter. This is immune response, which results in a) the production of antibodies, b) antibody-bearing cells or c) cell mediated hypersensitivity reaction (allergy). Each re-entry of the exogenous entities triggers enhanced production of the corresponding antibodies (booster reaction). The antibodies recognise their antigens, bind with them, and neutralise them, before they can cause harm to the individual or cause the disease specific to them. This is immunological defence. The antigen-antibody recognition is a highly specific phenomenon of biorecognition at the molecular level. Such a high degree of specificity is also found between the enzymes and their substrates, and lectins and their specific carbohydrates.

Immune response is a selective reaction of a mammalian body to substances that are foreign (exogenous) to it or those that the immune system identifies as foreign. The three important aspects are:

a) Memory: the primary response of the formation of the memory template at the first encounter,

b) Distinction between self- and non-self: distinction between the organism's endogenous proteins and those that are foreign (exogenous), and

c) Specificity: the secondary response of production of antibodies very specific to each foreign agent.

Other animals and even plants have defence mechanisms against diseases, but they are not identical to the mammalian immunological defences. For this reason, immunology is a mammal centred area of biology. Mice, rats, rabbits, dogs, horses and monkeys have been instrumental in the advancement of the field of immunology. Mammalian systems produce a highly specific antibody to each of the pathogens and even their different strains. The antibody recognises the antigen and binds with it forming an antigen-antibody complex, and neutralise the pathogen's potential to cause the disease. The antigen-antibody complex is scavenged by the body, mainly through the lymphatic system and often seen as pus, in dermal eruptions in the form of pustules. Pus, the fluid from pustules, contains serum, antigen-antibody complexes, expended white blood cells, dead and live pathogens, and debris of tissue.

## Components of the Immunological System and their Role

A large number of complex and interconnected components operate in the immune system. Understanding the role each one of them is essential to manipulating the system to our advantage.

### The Fluid Components

***Serum:*** The liquid part of blood, without the cells and the coagulating factors, but containing antigens and antibodies; it is the storehouse and means of transport of immunological components.

***Lymphatic system:*** is parallel to the blood conducting system and is constituted of the lymph, lymphocytes, lymph vessels, lymph nodes and lymph glands. The lymph is a watery, transparent or slightly yellow, fluid conducted through the lymph vessels. Lymph contains only one kind of cells, the lymphocytes, unlike blood that contains several different kinds of cells including lymphocytes. The blood and the lymphatic systems come into a sort of confluence in the lymph nodes and the tissues. The lymph cells that secrete lymph are aggregated into lymphatic tissue in the form of glands or occur in small groups of cells in different parts of the body.

### The Cells and Tissues

***Haemopoiesis (haematopoiesis):*** The formation of the cellular components of blood, originating very early in the yolk sac of the egg. In the foetus the liver performs this function and later the bone marrow takes it over and continues throughout life. Haemopoiesis originates with the stem cells in the bone marrow.

***Stem Cells:*** Stem cells are the basic cell type with potential to develop different cell components of the mammalian body system. Stem cells from the foetus are totipotent and can form almost any organ. The stem cells from the bone marrow are multipotent and form the cellular elements of both the blood and lymphatic systems in addition to the formation of new stem cells. The stem cells migrate to the thymus and differentiate into T-lymphocytes, in the microenvironment of the thymus.

***Erythrocytes:*** The enucleate discoid cells in the blood with membrane bound haemoglobin, to which oxygen binds reversibly (red blood corpuscles). Erythrocytes bear antigens on their surfaces that are responsible for the human blood groups in the ABO system. Blood group antigens also circulate in the blood and hence are responsible for the rejection of transplanted tissues/organs.

***Leucocytes:*** All the different kinds of cells in the blood (the so-called white blood corpuscles), including the lymphocytes, but with the exception of the erythrocytes.

***Lymphatic tissue/cells:*** As explained above, the lymphatic system is also composed of cells and tissue.

***Lymphocytes:*** The cells of the lymphatic system (lymphoid group) which play the main role in immune responses.

***Role of lymphocytes:*** The lymphocytes have an important role to play both in humoural and cell-mediated immunity. The lymphocytes re-circulate in the blood, lymph nodes, spleen and other tissues and back to blood by the lymphatic vessels.

When rats were depleted of lymphocytes, their ability to show the primary response to antigens or to reject skin grafts was very much impaired. Immunological responses were restored in these rats when lymphocytes from another rat were injected. This adequately shows the importance of lymphocytes in mounting immune response.

Kinds of lymphocytes:

a) T-lymphocytes, of four subclasses and the B-lymphocytes (T-cells and B-cells) basing on origin, and
b) Three kinds of lymphocytes, large, medium and small, basing on size.

When lymphocytes are incubated at 37°C for 24 h, the large and medium lymphocytes are killed. The remaining small lymphocytes can restore immune responses when injected into rats that were previously drained of lymphocytes.

The small lymphocytes are necessary for the primary response to an antigen and they can become

a) Antibody synthesising cells called plasma cells, or
b) Effector cells called lymphoblasts.

The lymphoblasts, along with blood group antigens, are responsible for immunological tissue rejection reactions in transplantations. The small lymphocytes also carry the memory of the first contact with an antigen. Without this memory mechanism, there can be no secondary response and so no immunological defence.

***Hybridoma:*** A synthetic cell line (such a myeloma cell and a spleen cell) that can grow in a culture indefinitely, at the same time producing antibodies.

Thymus: A gland lying behind the breastbone and extending up to the thyroid gland. The thymus is well developed in the infancy (about 40 g) and reaches its greatest size at about puberty (100 to 120 g) but is reduced by about 50 years of age (about 20 g), as it is progressively replaced by fatty tissue. So long as it is occurring, the thymus mediates the differentiation of the T-lymphocytes, which are concerned mostly with cell-mediated immunity. When the thymus was removed from mice at birth, they showed a decrease in lymphocyte count, their ability to reject tissue grafts was severely affected, their humoural antibody response was restricted and they soon died. When mice without the thymus were grown under germ free conditions, they survived showing that the ability to fight infection was impaired due to the removal of the thymus.

When mice were subjected to x-rays, their lymphocytes failed to multiply. When these mice were injected with bone marrow cells, their lymphocyte count normalised but not in mice without the thymus. These studies emphasise that the bone marrow cells develop into lymphocytes and that the thymus is necessary for this process.

Children with abnormalities of the thymus suffer from immunological disorders.

***Bursa of Fabricius:*** In birds, there is a recognisable lymphoid organ called the Bursa of Fabricius, which is similar in structure to the thymus. The Bursa is responsible for the production of B-lymphocytes that are involved in the humoural immunity.

When the Bursa was removed, chicken failed in humoural antibody synthesis but not in cell-mediated responses.

The lymphocytes that differentiate in the microenvironment of the Bursa are different from the T-lymphocytes and so are known as the B-lymphocytes. In man and other mammals, the bursa or its equivalent has not yet been identified. However, foetal liver and bone marrow cultures provided adequate evidence to consider that the B-lymphocytes in mammals differentiate in the microenvironment of the blood cell producing haemopoietic tissue of the bone marrow.

Distinguishing T- and B-cells: It is very difficult to distinguish between the T- and B-lymphocytes using a light microscope or even an electron microscope but certain tests ensure this. One of the common methods used to recognise human T-cells is to mix them with the red blood corpuscles of sheep when the two kinds of cells form rosettes (formations resembling roses). The B-cells are recognised by using fluorescent dyes along with anti-immunoglobulins (antithetic antibodies).

***Modified T- and B-cells:*** The populations of both the T- and B-cells are stimulated to proliferate and undergo morphological changes by antigens. The T-cells become lymphoblasts and participate in cell-mediate reactions. The B-cells become the plasma cells participating in the humoural antibody synthesis. There is co-operation between the two populations of lymphocytes. The mature plasma cell actively synthesises and secretes the antibody. There are no antibodies in, or secreted by, the T-lymphocytes.

***T-cell dependence of B-lymphocytes:*** Certain of the B-lymphocytes in mammals are dependent upon the T-lymphocytes for their function (T-cell dependent) while the others are independent of the T-lymphocytes (T-cell independent).

Monocytes, macrophages and phagocytes: Monocytes, originate from stem cells, have a single nucleus and develop into macrophages—the phagocytic cells, which engulf particulate matter, in a non-specific defence mechanism.

***Mast cells:*** Mast cells occur in the skin and epithelial layers. They contain histamine in the form of granules bound to membranes. Explosive de-granulation results in the release of histamine, which increases the permeability of the blood vessels, causing inflammatory reactions. Mast cells have a key role in allergy.

***Eosinophils:*** These are cells with granules in the cytoplasm (one kind of granulocytes), also known as polymorphonuclear leucocytes, stainable with the reddish biological stain eosine. The mast cells and eosinophils have an important role in allergy.

### The Molecules

***Antigen:*** a substance, usually a protein, that stimulates the immune system to produce a set of specific antibodies and that combines with an antibody specific to itself, at a specific binding site; differs from immunogen in that it is not involved in eliciting cellular response and in that it can complex with antibodies.

***Immunogen:*** a substance, usually a protein, that elicits a cellular immune response, and/or antibody production; differs from antigen in that it mainly elicits cellular response but does not complex with an antibody.

***Hapten:*** a low-molecular weight non-protein molecule which contains an antigenic determinant but which is not itself antigenic unless it complexes with an antigenic carrier, such as a protein; once an antibody is available, it can readily recognise the hapten, even without the carrier, and bind with it. To be antigenic, the hapten must bind to an exogenous protein carrier.

***Epitope:*** a part of a protein molecule that acts as an immunogenic/antigenic determinant, and so determines specificities; a macromolecule, such as a protein, may contain many different epitopes, each capable of stimulating the production of specific antibodies, each with a correspondingly specific binding site.

***Antibodies:*** Globulin (roughly spherical in shape and extractable in saline solutions) glycoproteins (proteins with a carbohydrate content ranging from 3 to 13%), produced by the immune system of an organism in response to exposure to a foreign molecule and characterised by its specific binding to a site, related to an epitope of that molecule; induced response proteins. The antibodies, like all proteins, are formed of chains of amino acids, which undergo very complex packing, giving the proteins a specific and functionally significant final shape (tertiary configuration), which determines most of the characteristics of the protein. As globulin proteins are involved in immune reactions, antibodies are called immunoglobulins (abbreviated to Ig).

***Antiserum:*** Blood serum containing antibodies arising out of immunisation or after an infectious disease.

***Production of antibodies:*** Antibodies are produced by the lymphocytes. The process of antibody production and immune response are complex and both the lymphatic and the blood systems are very closely involved.

***Autoantibodies:*** In certain pathological conditions, the thymus may produce antibodies to the body's own endogenous proteins (auto-antibodies), which complicates the immune system.

***Antithetic antibodies:*** Antibodies produced against antibodies; antithetic antibodies have properties similar to those of the antigens.

***Polyclonal antibodies:*** antibodies produced by molecules with several different antigenic determinants (epitopes) and/or several different cell populations.

***Monoclonal antibodies:*** antibodies produced against a single antigenic determinant (epitope) and/or by a single cell population; hence are very specific.

***Vaccine:*** An agent containing antigens/immunogens produced from killed, attenuated or lives pathogenic microorganisms, synthetic peptides, by recombinant organisms or DNA, used for stimulating the immune system of the recipient to produce specific antibodies providing active immunity and/or passive immunity.

### The Immunoglobulins

***Immunoglobulin (Ig):*** A protein molecule of the globulin-type, found in the serum or other body fluids and that possess antibody activity; there are five classes of immunoglobulins (IgA, IgD, IgE, IgG and IgM), based on antigenic and structural differences. In addition to these five classes, there are several subclasses (four in IgG) and other variants of Ig molecules.

***Classes of antibodies:*** There are five classes of immunoglobulins in the human system: Immunoglobulin G (the gammaglobulins; IgG), IgA, IgM, IgD and IgE.

***Molecular structure of the antibodies:*** The conventional model of the Ig molecules is a 'Y' shaped configuration, with two heavy chains and two light chains, with two open arms containing the antigen combining sites, which occur on both the light and the heavy chains. The two heavy chains are bound together by disulphide bonds. At any point, the molecule has two chain sections, parallel to each other.

The modern view of the structure of the Ig molecules is to look at them as containing series of regions activity called domains. Variable light, variable heavy, constant heavy 1,2,3 and constant light are the domains recognised on Ig molecules. The constant domains provide for the identity of the molecules and the variable regions are responsible for the diversity in the specificity of the antibodies.

Ig molecules may occur as monomers (IgG and IgA), dimers (IgA) or pentamers (IgM). Linking monomers by J chains forms higher configurations.

***Immunoglobulin A (IgA):*** With a molecular weight of about 1,60,000, IgA molecules are only slightly heavier than the IgG molecules but they can form aggregates of higher molecular weights. IgA are about 13% of the total Ig with a concentration of 1.4 to 4 mg/ml in the normal serum. The IgA are the major Ig in the serum and mucous secretions, such as saliva, tears, nasal fluids, sweat, lung and the gastrointestinal tract. They defend the exposed external surfaces of the body against the attack of microorganisms. IgA antibodies seem to inhibit adherence of the microorganisms to the surface of the mucosal cells and thus prevent their entry into the body tissues. IgA molecules differ from the other Ig classes in having three disulphide bonds holding the two heavy chains, instead of two bonds in the others.

***Immunoglobulin M (IgM):*** The IgM molecules are the heaviest of all Ig. They have a molecular weight of 900,000 and so are often known as the macroglobulins. They form about 6% of the total Ig and occur in a concentration of 0.5 to 2% of the normal serum. IgM are very efficient agglutinators of bacterial cells and are effective cytolytic agents. They form the most immediate and effective first line defence against bacteraemia. Since they appear in response to infection they are mostly confined to the blood stream. The anti-A and anti-B haemagglutinins and many antimicrobial antibodies as well as typoid exotoxin antibodies are all IgM. During the course of evolution of Ig, IgM seem to have appeared earliest.

***Immunoglobulin G (IgG):*** IgG molecules are the lightest of all the Ig and have a molecular weight of about 1,50,000 and about 3% carbohydrate content. They form about 80% of the total Ig of the human body. In the normal serum their concentration ranges from 8 to 16 mg/ml. These are the most abundant component of Ig in the body fluids particularly the blood vessels where they combat microorganisms and their toxins.

IgG are the only antibody that can get across the placenta and so provide the major line of defence during the first few weeks of the life of a foetus. IgG also diffuse very readily from the blood vessels into the body spaces. When IgG molecules attach to microorganisms, the susceptibility of the latter for phagocytosis increases. In a germ free environment, the IgG concentration of the serum is very low and increases with infection. IgG are the major antibody synthesised during the secondary response, their synthesis being entirely governed by the antigenic situation.

All the IgG molecules are seemingly identical. The most fascinating thing is that there are an infinitesimal number of antigens, with each pathogenic organism producing several of them. During the course of our lifetime we develop immunity against a very large number of infections, some on a long-term basis and some ephemeral but repeated infection renewing our ability to combat the disease.

The key to understanding this versatility of the IgG molecule lies in the fact that the IgG molecule has a part that is invariable and this gives the basic characteristics for it to function as an antibody. Another part of the IgG molecule is variable in its amino acid content and sequence and this gives the molecule the ability to be a specific antibody against a particular antigen. This is nothing surprising. Almost all proteins have variable and invariable regions.

***Immunoglobulin E (IgE):*** The molecular weight of IgE is about 200,000 and they form only 0.002% of the total Ig with a serum concentration of 17 to 450 ng/ml. IgE protect the external mucosal surfaces of the body through plasma factors. Pathogens crossing the IgA line combine with IgE molecules specific to them. This results in the release of amines (eg. histamine) that increase the permeability of the blood vessels causing the symptoms of allergy. The release of amines is due to a degranulation of the mast cells. The level of IgE is raised during parasitic infections but the importance of IgE lies with atopic allergy.

***Immunoglobulin D (IgD):*** IgD have a molecular weight of about 1,85,000 and form only about 1% of the total Ig. They occur at a concentration of 0 to 0.4% of the normal serum. They are present only on the surface of the lymphocytes along with IgM. The IgD are susceptible to enzyme degradation and so have a very short life span (2.8 days) in the plasma. IgD have the highest carbohydrate content (13%) of all Ig. The exact function of IgD is not understood.

## Types of Immunological Reactions

***Agglutination:*** an immunological or chemical reaction leading to the aggregation of particulate matter such as bacteria, erythrocytes or other cells, or synthetic particles such as plastic beads coated with antigens or antibodies. Precipitin reaction: When an antigen and its antibody are brought together in solution, a precipitate is formed due to the binding of the antigen and the antibody. If unrelated antigen and antibody are brought together no binding occurs and hence no precipitate is formed. Antigen-antibody binding occurs when they come together in the blood stream or in the tissues. Precipitation occurs because the antigen-antibody complexes form a three-dimensional lattice. Precipitin reactions are a very useful tool in several areas of biological research.

In the case of both antigen-antibody and enzyme-substrate affinity, there is a complementarity of the molecular shape between the antigen/enzyme and the antibody/substrate and the fit is exact like that of a key in its lock. In semisolid media, such as bacto agar, the precipitin reaction results in the formation of lines called precipitin lines. Such reactions are studied by Ouchterlony's double diffusion method, where the antigen and the antibody diffuse towards each other from two spaced wells cut in semisolid agar. This method provides only qualitative data. A variant of this method is single radial diffusion, which helps to quantify the antigen with reference to the antibody.

Basis of recognition of the antigen by the antibody and their binding: The overall physical configuration of the antigen seems to be more important than its chemical structure which means that the antigen is recognised by the three-dimensional shape of its outer electron cloud. Chemical composition and reactivity are less important.

***Binding site:*** a specific region in a molecular entity, such as an antigen, that is capable of entering into a stabilising interaction with another molecular entity, such as the corresponding antibodies.

***Forces of antigen-antibody binding:*** One or more of the following forces appear to be involved in antigen-antibody binding:

electrostatic forces, hydrogen bonding, hydrophobic (water repulsion) forces and Van der Waals attractions between molecules. What is surprising is that the very same forces also operate between unrelated proteins or other macromolecules in normal chemical reactions.

### Types of Immunity

***Humoural immunity:*** When microorganisms enter the body, antibodies are synthesised and released into the blood and other body fluids. These antibodies circulate throughout the body. The free antibodies coat the cells of the organism and enhance their phagocytosis and also neutralise the toxins released by the organisms. This type of immune response is called the humoural immunity.

***Cell-mediated immunity:*** In response to the presence of antigens, the body produces lymphocytes with antibodies or antibody-like molecules on their surface. This is cell-mediate immunity, which offers protection, particularly against organisms, which live and multiply within the host cells. Tubercle bacteria, small poxvirus, etc., are subject to the action by cell-bound antibodies.

***Acquired immunity:*** Not all antibodies are synthesised in the body. Some are pre-natal acquisitions from the mother through the placenta and some are post-natal through breast-feeding. These constitute acquired immunity. Immunity is also acquired through one's own body's experience gained on encountering a pathogen.

***Specific and non-specific defence:*** Immunological defence is specific to particular pathogens, and even to their strains. This is specific defence. Mammals have also a non-specific defence mechanism. For example, the macrophages, that are associated with the lumenal side of the walls of the blood vessels and the connective tissue, physically engulf cells of pathogens or complexes of proteins, to remove them from the system.

### Body's Immunological Reaction

***Allergy:*** A hypersensitivity reaction of the body to antigens. In a sense it is the immunological system that has gone wrong. An allergen is an antigen that stimulates the production of IgE antibodies, although low titres of IgG molecules are also formed. The IgE antibodies bind to mast cells resulting in the 'explosion' of the mast cells leading to the release of histamine that triggers an inflammatory response in the skin, mucosa or epithelial cells, a syndrome termed allergy.

***Allergen:*** an antigen that can induce an allergic reaction, thorough eliciting IgE antibodies. Some allergens are haptens, as for example parthenin which is a sesquiterpene lactone. Some haptens bind to endogenous proteins in the individual, because of which IgG antibodies cannot be produced against such a hapten-carrier complex, a situation that makes it almost impossible to treat the patient through immunisation.

***Anaphylaxis:*** a sudden and severe form of IgE based reaction, occurring on the second encounter with the allergen (antigen) that can be fatal. Penicillin may induce a severe anaphylactic reaction in some individuals sensitive to it. In fact, purified or synthetic penicillin does not cause anaphylaxis, but the impurities in biologically produced penicillin or protein compounds such as procain added to penicillin injection, are responsible for the reaction.

***Inflammatory response:*** This is the body's reaction to injury or infection/antigens, in the form of a syndrome constituted of swelling, redness (erythema) and heat (collectively called inflammation), in the affected part of the body. Inflammation controls the spread of infection. Uncontrolled inflammation causes tissue damage.

### *Inhibition of Immune Response*

Antigen-antibody completing, phagocytosis, inflammation during immune response, etc., all lead to tissue damage. Uncontrolled immune reactions can be dangerous to us. There are some factors that inhibit immune response and some situations where immune responses fail to materialise.

a) Antibody suppressor cells and/or factors are present in our serum or tissues. Prostaglandins, the compounds secreted by the organs of the human body into the blood stream to perform various functions, such as muscle contraction, may also inhibit immune reactions. This is the body's way of controlling immune reactions to minimise tissue damage.

b) While our body prepares for extensive warfare, the pathogens themselves would not be idle. A number of immunosuppressive agents like lipopolysaccharides, lipoteichoic acid, dextran, Levan, etc., are produced by bacteria. They also produce proteinases that denature some Ig classes, particularly IgA. Modulation of immune responses both by the host and the pathogen ultimately regulates the host-pathogen interaction and the development of disease.

c) The antibodies can be defeated in their function by slight changes in the chemical (and consequently physical) structure of the antigen. This happens with the antigens of viruses and bacteria, which grow very rapidly and develop into new strains through mutations and other evolutionary processes depending upon several conditions, particularly environmental. Under these conditions, the host's immune defences become inadequate. For example, we never seem to acquire immunity from colds. In fact, we do get immunised to a particular strain of cold causing virus of a given time but in no time the virus modifies itself in some minute way and we have no immediate defence against this modified version of the virus. Antibody production against the ever-changing organisms is a race between the host and the pathogen.

Clinical suppression of immune response: At the time of tissue and organ transplantation from one individual to another, the immune system of the recipient's body, especially the lymphoblast component triggers the production of antibodies against the antigens in the tissue/organ of the donor, which results in the rejection of the transplant. This is also because of the presence of blood group antigens in tissues, in addition to principally being on the surface of the erythrocytes. In order to prevent this situation, the immune responses are deliberately suppressed by the use certain drugs such as azathioprine, cyclophosphamide and cyclosporin A, prior to transplantation. Such drugs are also used in the event of autoimmune diseases, like rheumatoidal arthritis.

### *Secretors and Non-secretors*

All of us secrete antigens and antibodies in our body fluids such as sweat, tears, saliva, semen, etc., to some degree or the other. For example, the IgA in saliva serves as the first line of defence of the oral route. However, a certain percentage of human populations secrete antigens and antibodies in high titres and are called secretors, the other group being non-secretors. The status of an individual as a secretor is genetically determined and offers certain advantages to the secretors in terms of body hygiene. Their surface and first line defences are quite high compared to those of the non-secretors. The frequency of secretor *vs* non-secretor alleles in different human populations is of interest to the population geneticist. The status of an individual as a secretor (or non-secretor) is easily determined by the use of appropriate lectins, a class of proteins that can recognise and bind to cell surface carbohydrates, resulting in agglutination of cells, such as erythrocytes and lymphocytes.

### Vaccines

A number of diseases are caused by micro-organisms. For many bacterial diseases, and some fungal diseases, there are antibiotics, produced by other micro-organisms. There are also chemical agents produced by higher plants that can control microbial pathogens. But there are very few means of fighting viral diseases. Even when there are therapeutic agents to control several microbial diseases, in course of time the pathogen acquires resistance, making the therapeutic agent ineffective. In this eternal race between the pathogen and the pathologist, the pathogens seem to be always a step ahead. Against this background, production of vaccines against the microbial pathogens, and more particularly the pathogenic viruses, in order to immunise the susceptible populations, is a safe and more certain recourse. More importantly, the immunological approach is both preventive and curative, while other means are only curative. Advances in immunology and biotechnology have made it now possible to produce immunological agents to afford protection from diseases to large numbers of people. This area is immunotechnology, an arm of biotechnology.

A vaccine is an agent, sourced from the pathogen, and is deliberately introduced into the mammalian system in order to impart a 'memory' of the pathogen or its pathogenic component. The memory is imparted on the first contact of the vaccine with the mammalian immune system. Vaccine usually contains the modified pathogenic organism or a protein or a low molecular weight non-protein compound (hapten) conjugated with the protein, obtained from the pathogen. Vaccines contain antigens (that elicit the production of antibodies), or immunogens (that trigger the cellular component of immune response). In the event of an encounter with the corresponding antibodies, only the antigens can bind with the antibodies, and form an antigen-antibody complex that neutralises the harmful effects of the antigens or the organisms that produce them. Since vaccines employ a part of the chemical machinery of the organisms themselves, pathogens cannot easily acquire resistance to vaccines, as they do for antibiotics and chemical therapeutic agents.

### Vaccination/Immunisation

The process of the deliberate introduction of a vaccine into the organism is vaccination, for which the term inoculation is also often used. Since vaccination immunises the organism, the process is also called immunisation. When an organism is vaccinated, the immune system is readied to show an immune response by way producing antibodies against the pathogen, in the event of a second encounter with the pathogen, basing on the memory imparted by the vaccine used for the first encounter.

### Immunotherapy

Immunotherapy differs from vaccination in that in the former antibodies isolated from an immunised organism (polyclonal or monoclonal immunoglobulin antibodies or cytokines) is used to cure the patient. Immunotherapy becomes essential when there is no time to prepare the patient through vaccination or when the patient is physically and/or physiologically not competent to respond to vaccination.

### Composition of Vaccines

Vaccines are suspensions, in saline or buffered saline, of weakened pathogenic organisms or their fractions or the proteins they secrete, which have the potential to cause a disease. A virulent organism cannot be used as a vaccine.

### Adjuvants

Antigens often need to be coupled with an adjuvant, which is a compound that holds the antigen and releases it slowly over a longer period of time. The most commonly used immunological adjuvant in experimental systems is Freund's Complete Adjuvant, which contains mineral oil and heat killed mycobacteria. The bacteria are intended to heighten immunological response, but may produce hypersensitivity in many patients. The mineral oil also may prove to be harmful. Hence FCA is not normally used in human immunisation schedules. Aluminium hydroxide is human safe but is a poor adjuvant. Some plant saponins are now projected as efficient and human safe adjuvants. There are some effective and safe synthetic adjuvants, but their composition is a trade secret.

### Types of Vaccines

a) Inactivated vaccines: The pathogen is killed using heat or formalin, as for example, typhoid or Salk poliomyelitis vaccines.

b) Attenuated vaccines: The pathogen is weakened (attenuated) by aging or altering growth conditions, but is alive, as in the case of measles, mumps and rubella vaccines. There is some risk of the concerned virus becoming virulent.

c) Avirulent organisms: A non-pathogenic strain of a pathogenic organism is used as a vaccine, as in BCG (Bacillus Calmette Guerin) vaccine against *Mycobacterium tuberculosis,* the tuberculosis bacterium.

d) Toxoids: The toxin from the pathogen is used as an antigen to produce the vaccine. The severity of the toxicity of the antigen is reduced by treating it with aluminium salts while preparing the 'toxoid', as in the case of diphtheria and tetanus.

e) Acellular vaccines: Only the antigenic component of the organism is used instead of the whole organism, as in Haemophilus influenza B vaccine.

f) Subunit vaccines: Genetic engineering techniques have now made it possible to use as a vaccine only a part of an organism that is adequate to stimulate the immune response. An appropriate segment of genetic material is isolated from the pathogens and introduced into bacteria or yeasts, to transcribe and translate the inserted foreign DNA. The product is used as a vaccine, as in the case of Hepatitis B vaccine. These vaccines cannot cause the disease even in patients whose immunological system is impaired (immunocompromised) patients.

g) DNA vaccines: Described as the third vaccine revolution, DNA vaccines are an offshoot of gene therapy. Selected segments of DNA, when introduced into the patients system synthesise and deliver proteins that are needed to replace the defective enzyme system or tag a cell for destruction. Viruses or lipid vehicles are used to deliver the DNA into the cells. This recent technology is being tried to produce vaccines against HIV, by a direct injection of plasmid borne DNA.

### *Herd Immunity*

Use of vaccines to prevent disease in communities is herd immunity, which affords protection by decreasing the number of susceptible people in a community, with time. This basically constitutes mass immunisation. Polio vaccination programmes now target an enormous number of children throughout the world, to eradicate polio, as was done for smallpox earlier.

### *Booster Doses*

The effectiveness of certain vaccines is life long as for example of smallpox, measles, mumps and rubella. Attenuated vaccines normally afford life long immunity. But in the case of certain others, the effectiveness is short lived and the immune system needs to be re-educated through periodical booster doses. The vaccine is administered one or more times, with appropriate time gaps, after the initial vaccination, to boost to the immune system to produce adequate quantities of antibodies against the intended pathogen. Toxoid vaccines require a booster every ten years or so. Booster doses are also needed in case of inactivated or acellular vaccines, which are very safe, as they cannot cause the disease.

### *Multiple Vaccines*

While most vaccines contain antigens of a single pathogen, there is a practice of multiple vaccines, which combine antigens of more than one pathogen. For example, diphtheria, tetanus and pertussis are administered together as DTP vaccination.

### *Vaccine Administration*

Vaccines are administered, as injections (DTP), or dermally (smallpox, anthrax), or orally (polio) or as a nasal spray (influenza virus).

### *Edible Vaccines*

Now transgenic plants are being developed through genetic engineering techniques, where the vaccine is synthesised in the edible part of a food plant (edible vaccines). Transgenic bananas, melons, and tomatoes are choice candidates for carrying edible subunit vaccines, as for example against rabies. The obvious advantage is the ease of transportation and storage of the vaccine bearing material and administration without technical support. Conventional vaccination programmes in many countries are seriously handicapped due to a lack of equipment for storage and transport of the vaccines and the shortage of paramedical staff to administer the vaccines.

### *Safety of Vaccines*

By and large vaccination programmes have proven to be reasonably safe for the human populations. However, at certain times complications may arise mostly due to an incorrect handling of the vaccines and/or vaccination or due to individual metabolic deficiencies. In spite of all that is adverse in vaccination, immunisation is one of the most efficient means of disease prevention, particularly in large sections of the human population. In the case of HIV and epidemic diseases and even cancer, immunisation is probably the only hope.

## Labelling and Traceability of Genetically Engineered Foods

### *The Indian Context*

Labelling commercialized products dates back to the times of the Roman Empire. Over the centuries the product label has become a means of product identification, establishing Intellectual Property Rights, brand presence and market position, by the manufacturer. For the authorities labeling is a tool of protecting consumer interests and of tracing the antecedents of the product, to check any regulatory violations by the manufacturer and the marketing outlets. Over the years, labeling, a simple means of providing information to the consumer to make educated choices, became a very complex and contentious issue, in almost every sphere. Plants and animals, that are source of food and feed, are genetically engineered (GE) to enhance flavor, quality and yield, to increase nutrients, and to improve resistance to pests and diseases. GE plants and animals have turned out to be one of the most controversial issues today, as the opponents of technology have raised questions of safety and the ethics behind such technology. Labelling and traceability regulations for non-GE food products have existed for decades in almost every country.

All countries that are involved in the development, cultivation and marketing of GE crops, have now a regulatory frame work for consumer safety of food derived from GE plants and animals. Procedures and regulations of testing GE foods for different safety parameters and mandatory labeling practices, to facilitate traceability of a food product to its genetic and production source, have unfortunately become international controversies. However, a uniform international policy is essential to smoothen international trade in such products.

The Food and Drug Authority of the US is satisfied with the establishment of Substantial Equivalence between a GE product and its isogenic, and does not have mandatory labeling regulations. However, labels and symbols for GE foods, are used voluntarily in the US, to indicate if a product is 'GM free'. The tag 'Identity Preserved' is used to refer to the record of a product's specific traits through the entire process from the crop to the product. The European Union has detailed regulatory procedures, to ensure the safety of GE food products to the consumers and labeling to provide product information to both the user and the authorities. The EU's regulations are so stringent that they sparked protests from the product developers and raised doubts whether any GE product will at all qualify to be considered safe under these regulations.

Codex Alimentarius Commission (CAC) is the international organization established in 1963, jointly by the FAO and WHO, under the Food Standards Programme. CAC is an inter-governmental body whose membership is open to all Member Nations and Associate Members of FAO/WHO, and currently comprises of over 165 countries. International non-governmental organizations, such as consumer, academic or industry bodies, may attend Codex meetings as observers.

The objectives of CAC are, protecting health of the consumers, ensuring fair trade practices in the food trade, and promoting co-ordination of all food standards work undertaken by international governmental and non-governmental organizations. The Codex has framed detailed policy guidelines to establish safety of products of modern biotechnology and foods derived through GE plants and micro-organisms. However, a number of gray and conflicting areas dog consensus.

The Cartagena Protocol is not relevant to food safety and labeling issues as it is primarily concerned with international trans-boundary movement of GE plants and animals and neither with foods nor their labelling.

Some of the much debated questions are:

- What kind of label would serve the objective of consumer and regulatory needs?
- Should it be linguistic or symbolic or a combination of the two?
- What a label should include?
- Should it be mandatory or voluntary?
- What is threshold of GE content in a food product?
- Should refined products derived from GE plants and animals, but do not contain any GE component be labelled?
- What kind of scientific procedures be adopted in the qualitative and quantitative analysis of GE component in foods?

In multilingual countries, the choice of languages in providing label information is also a serious emotionally charged question. At the end of all this time consuming and expensive exercise, the question that remains is, 'how many consumers actually read the label information and are guided by it in their choice?'.

Many countries that introduced strict labelling and import regulations for GE products have already been threatened under WTO trade agreements, for example, the EU by the US. In India, the 'Prevention of Food and Adulteration Act' of 1954 was aimed to prevent the adulteration and misbranding of food. The label was required to indicate the name or description of the food, the name and business address of the manufacturer or importer, the net weight of the food, ingredients, batch/code number in Hindi or English (or regional languages), the month and year of manufacture, packing and expiry and If preservatives, coloring agents, antioxidants, or vitamins have been added to the food. Recently, a symbolic label was introduced to distinguish entirely vegetarian foods from foods that have a non-vegetarian component. India should now consider framing internationally compatible regulations on labelling and traceability of GE food products. The issues have to be addressed case-by-case basis. The only GE crop now grown in India is Bt cotton and it has no serious implications for use as food. Though there are several GE crops in development, both in the public and private sectors, no GE food product is likely to emerge in India in this decade. Framing rules for labelling and traceability of GE foods at this point of time would be without a focus.

However, it would immensely benefit, if a working group reviews the provisions of international instruments and the regulations adopted in different countries and draft a basic policy, with a provision to suitably modify as and when a GE product would reach the market.

The product developers and stakeholders should be involved in the process in order to arrive at consensus. A move in this direction has recently been made by the Indian Council of Medical Research, Government of India, New Delhi. This process will be a little easier if the working group has authentic information on the probable datelines of release of different GE products, both in the public and private sectors.

Summary of talk delivered at the International Conference on 'Foods derived from genetically modified crops: issues for consumers, regulators and scientists' at New Delhi, September 26-27, 2005.

## Fermentation Biotechnology

'Biotechnology', the short form of Biological technology, defies precise definition.

The term biotechnology came into general use in the mid 1970s, gradually superseding the more ambiguous 'bioengineering', which was variously used, to describe chemical engineering processes using organisms and/or their products, particularly fermenter design, control, product recovery and purification. Most scientists agree that all processes that utilize biological organisms constitute biotechnology, but what is disputed is which processes do not (Crafts-Lighty, 1983). Definitions of technology too vary from the simple 'applied science' to 'the scientific study of the practical or industrial arts' (Crafts-Lighty, 1983).

In consequence of the disagreements, there are at least ten different definitions of biotechnology, each qualified according to the context of use. The following are among the more widely used definitions of biotechnology:

a) "The application of biological organisms, systems or processes to manufacturing and service industry" (HMSO, 1980).

b) "The application of scientific and engineering principles to the processing of materials by biological agents to provide goods and services" (Bull et al., 1982). Agents include a wide range of biological substances, such as enzymes, as well as whole cells or multicellular organisms. 'Goods and services' covers processes such as waste and water treatment.

c) "The integrated use of biochemistry, microbiology and chemical engineering to exploit plant materials and genetic resources for the production of specific products and services" (Mantell, 1989).

Many definitions are rather vague about the nature of the organism or agents involved in biotechnology. It is argued that systematic farming of plants and animals for food and fuel also falls within these definitions.

Ancient fermented food processes, such as making bread, wine, cheese, curds, *idli, dosa,* etc., some of which are some 6,000 yr old, and developed long before man had any knowledge of the existence of the micro-organisms involved, also genuinely constitute biotechnology. However, for the sake of convenience, many people exclude these traditional processes from the realm of biotechnology. Conventional agriculture is a well-developed industry in its own right, but in practice, this is not included in biotechnology. Aspects of 'modern biotechnology' may have significant effects on 'traditional biotechnology'. Genetic manipulation to improve brewing and baking yeasts or to introduce new characteristics in crops, biological control of plant pests, and new methods of diagnosing and preventing plant, human and animal disease, are all now realisable. Whatever the definition, experimental production of new varieties of organisms, is one of the important objectives of biotechnology.

In simpler words, biotechnology means the industry-scale use of organisms and/or their products. Now biotechnology virtually includes the scientific, technological and commercial aspects of almost every area of human welfare from agricultural production to pollution control.

### *Fermentation*

Fermentation technology is the oldest of all biotechnological processes. The term is derived from the Latin verb *fevere*, to boil—the appearance of fruit extracts or malted grain acted upon by yeast, during the production of alcohol. Fermentation is a process of chemical change caused by organisms or their products, usually producing effervescence and heat. Microbiologists consider fermentation as 'any process for the production of a product by means of mass culture of micro-organisms'.

Biochemists consider fermentation as 'an energy-generating process in which organic compounds act both as electron donors and acceptors'; hence fermentation is 'an anaerobic process where energy is produced without the participation of oxygen or other inorganic electron acceptors'. In biotechnology, the microbiological concept is widely used.

## Microbial Growth

### *Requirements for Artificial Culture*

The growth of organisms involves complex energy based processes. The rate of growth of micro-organisms is dependent upon several culture conditions, which should provide for the energy required for various chemical reactions. The production of a specific compound requires very precise cultural conditions at a particular growth rate.

Many systems now operate under computer control. The rate of growth of micro-organisms and hence the synthesis of various chemical compounds under artificial culture, requires organism specific chemical compounds as the growth (nutrient) medium.

The kinds and relative concentrations of the ingredients of the medium, the pH, temperature, purity of the cultured organism, etc., influence microbial growth and hence the production of biomass (the total mass of cells or the organism being cultured), and the synthesis of various compounds.

***Table:*** *Nutrient sources for industrial fermentation*

| ***Nutrient Raw material*** | |
|---|---|
| Carbon source | |
| Glucose | Corn sugar, Starch, Cellulose |
| Sucrose | Sugarcane, Sugar beet molasses |
| Lactose | Milk whey |
| Fats | Vegetable oils |
| Hydrocarbons | Petroleum fractions |
| ***Nitrogen source*** | |
| Protein | Soybean meal, Cornsteep liquor, Distillers' solubles |
| Ammonia | Pure ammonia or ammonium salts |
| Nitrate | Nitrate salts |
| Nitrogen | Air |
| Phosphorous source | Phosphate salts |

## Phases of Microbial Growth

When a particular organism is introduced into a selected growth medium, the medium is inoculated with the particular organism. Growth of the inoculum does not occur immediately, but takes a little while. This is the period of adaptation, called the lag phase.

Following the lag phase, the rate of growth of the organism steadily increases, for a certain period—this period is the log or exponential phase. After a certain time of exponential phase, the rate of growth slows down, due to the continuously falling concentrations of nutrients and/or a continuously increasing (accumulating) concentrations of toxic substances. This phase, where the increase of the rate of growth is checked, is the deceleration phase.

After the deceleration phase, growth ceases and the culture enters a stationary phase or a steady state. The biomass remains constant, except when certain accumulated chemicals in the culture lyse the cells (chemolysis). Unless other micro-organisms contaminate the culture, the chemical constitution remains unchanged. Mutation of the organism in the culture can also be a source of contamination, called internal contamination.

### *Fermenters and Bioreactors*

A fermenter is the set up to carry out the process of fermentation. The fermenters vary from laboratory experimental models of one or two litres capacity, to industrial models of several hundred litres capacity, which refers to the volume of the main fermenting vessel. A bioreactor differs from a fermenter in that the former is used for the mass culture of plant or animal cells, instead of micro-organisms. The chemical compounds synthesised by these cultured cells, such as therapeutic agents, can be extracted easily from the cell biomass. The design engineering and operational parameters of both fermenters and bioreactors are identical. With the involvement of micro-organisms as elicitors in some situations, the distinction between the two concepts is being gradually obliterated.

### *Design of Industrial Fermentation Process*

The fermentation process requires the following:

a) a pure culture of the chosen organism, in sufficient quantity and in the correct physiological state;

b) sterilised, carefully composed medium for growth of the organism;

c) a seed fermenter, a mini-model of production fermenter to develop an inoculum to initiate the process in the main fermenter;

d) a production fermenter, the functional large model; and

e) equipment for i) drawing the culture medium in steady state, ii) cell separation, iii) collection of cell free supernatant, iv) product purification, and v) effluent treatment. Items a) to c) above constitute the upstream and e) constitutes the downstream, of the fermentation process.

Fermenters/bioreactors are equipped with an aerator to supply oxygen in aerobic processes, a stirrer to keep the concentration of the medium uniform, and a thermostat to regulate temperature, a pH detector and similar control devices.

## Types of Culture Systems

### *Batch Processing or Culture*

At about the onset of the stationary phase, the culture is disbanded for the recovery of its biomass (cells, organism) or the compounds that accumulated in the medium (alcohol, amino acids), and a new batch is set up. This is batch processing or batch culture. The best advantage of batch processing is the optimum levels of product recovery. The disadvantages are the wastage of unused nutrients, the peaked input of labour and the time lost between batches.

### Continuous Processing or Culture

The culture medium may be designed such that growth is limited by the availability of one or two components of the medium. When the initial quantity of this component is exhausted, growth ceases and a steady state is reached, but growth is renewed by the addition of the limiting component. A certain amount of the whole culture medium (aliquot) can also be added periodically, at the time when steady state sets in. The addition of nutrients will increase the volume of the medium in the fermentation vessel. It is so arranged that the increased volume will drain off as an overflow, which is collected and used for recovery of products. At each step of addition of the medium, the medium becomes dilute both in terms of the concentration of the biomass and the products. New growth, stimulated by the added medium, will increase the biomass and the products, till another steady state sets in; and another aliquot of medium will reverse the process.

This is continuous culture or processing. Since the growth of the organism is controlled by the availability of growth limiting chemical component of the medium, this system is called a chemostat. The rate at which aliquots are added is the dilution rate that is in effect the factor that dictates the rate of growth.

The events in a continuous culture are:

a) the growth rate of cells will be less than the dilution rate and they will be washed out of the vessel at a rate greater than they are being produced, resulting in a decrease of biomass concentration both within the vessel and in the overflow;

b) the substrate concentration in the vessel will rise because fewer cells are left in the vessel to consume it;

c) the increased substrate concentration in the vessel will result in the cells growing at a rate greater than the dilution rate and biomass concentration will increase; and

d) the steady state will be re-established.

Hence, a chemostat is a nutrient limited self-balancing culture system, which may be maintained in a steady state over a wide range of sub-maximum specific growth rates. The continuous processing offers the most control over the growth of cells. Commercial adaptation of continuous processing is confined to biomass production, and to a limited extent to the production of potable and industrial alcohol. The steady state of continuous processing is advantageous as the system is far easier to control. During batch processing, heat output, acid or alkali production, and oxygen consumption will range from very low rates at the start to very high rates during the late exponential phase.

The control of the environmental factors of the system becomes difficult. In the continuous processing, the rates of consumption of nutrients and those of the output chemicals are maintainable at optimal levels. Besides, the labour demand is also more uniform. Continuous processing may suffer from contamination, both from within and outside. The fermenter design, along with strict operational control, should actually take care of this problem.

The production of growth associated products like ethanol is more efficient in continuous processing, particularly for industrial use. Continuous culturing is highly selective and favours the propagation of the best-adapted organism in culture. A commercial organism is highly mutated such that it will produce very high amounts of the desired product. But physiologically such strains are inefficient and give way in culture to inferior producers—a kind of contamination from within.

### Fed-batch Culture or Processing

In the fed-batch system, a fresh aliquot of the medium is continuously or periodically added, without the removal of the culture fluid. The fermenter is designed to accommodate the increasing volumes. The system is always at a quasi-steady state. Fed-batch achieved some appreciable degree of process and product control.

A low but constantly replenished medium has the following advantages:

a) maintaining conditions in the culture within the aeration capacity of the fermenter;

b) removing the repressive effects of medium components such as rapidly used carbon and nitrogen sources and phosphate;

c) avoiding the toxic effects of a medium component; and

d) providing limiting level of a required nutrient for an auxotrophic strain.

Production of baker's yeast is mostly by fed-batch culture, where biomass is the desired product. Diluting the culture with a batch of fresh medium prevents the production of ethanol, at the expense of biomass; the moment traces of ethanol were detected in the exhaust gas. The production of penicillin, a secondary metabolite, is also by fed-batch method.

Penicillin process has two stages: an initial growth phase followed by the production phase called the 'idiophase'. The culture is maintained at low levels of biomass and phenyl acetic acid, the precursor of penicillin, is fed into the fermenter continuously, but at a low rate, as the precursor is toxic to the organism at higher concentrations.

## Products of Fermentation Processes

The growth of micro-organisms or other cells results in a wide range of products. Each culture operation has one or few set objectives. The process has to be monitored carefully and continuously, to maintain the precise conditions needed and recover optimum levels of products. Accordingly, fermentation processes aim at one or more of the following:

a) production of cells (biomass) such as yeasts;
b) extraction of metabolic products such amino acids, proteins (including enzymes), vitamins, alcohol, etc., for human and/or animal consumption or industrial use such as fertiliser production;
c) modification of compounds (through the mediation of elicitors or through biotransformation); and
d) production of recombinant products.

### *Microbial Biomass*

Microbial biomass is produced commercially as single cell protein (SCP) using such unicellular algae as species of *Chlorella* or *Spirulina* for human or animal consumption, or viable yeast cells needed for the baking industry, which was also used as human feed at one time. Bacterial biomass is used as animal feed. The biomass of *Fusarium graminearum* is also produced, for a similar use.

### *Microbial Metabolites*

***Primary Metabolites:*** During the log or exponential phase organisms produce a variety of substances that are essential for their growth, such as nucleotides, nucleic acids, amino acids, proteins, carbohydrates, lipids, etc., or by- products of energy yielding metabolism such as ethanol, acetone, butanol, etc.

This phase is described as the tropophase, and the products are usually called primary metabolites.

***Table:*** *Examples of commercially produced primary metabolites*

| | | |
|---|---|---|
| Ethanol | *Saccharomyces cerevisiae* | alcoholic beverages |
| | *Kluyveromyces fragilis* | |
| Citric acid | *Aspergillus niger* | food industry |
| Acetone and | *Clostridium* | |
| butanol | *acetobutyricum* | solvents |
| Lysine | *Corynebacterium* | nutritional additive |
| Glutamic acid | *glutamacium* | flavour enhancer |
| Riboflavin | *Ashbya gossipii* | nutritional |
| | *Eremothecium ashbyi* | |
| Vitamin B12 | *Pseudomonas denitrificans* | nutritional |
| | *Propionibacterium shermanii* | |
| Dextran | *Leuconostoc mesenteroides* | industrial |
| Xanthan gum | *Xanthomonas campestris* | industrial |

***Secondary metabolites:*** Organisms produce a number of products, other than the primary metabolites.

The phase, during which products that have no obvious role in metabolism of the culture organisms are produced, is called the idiophase, and the products are called secondary metabolites.

In reality, the distinction between the primary and secondary metabolites is not a straightjacket situation.

Many secondary metabolites are produced from intermediates and end products of secondary metabolism. Some like those of the Enterobacteriaceae do not undergo secondary metabolism.

***Table:*** *Examples of commercially produced secondary metabolites*

| Metabolite | Species | Significance |
|---|---|---|
| Penicillin | *Penicillium chrysogenum* | antibiotic |
| Erythromycin | *Streptomyces erythreus* | antibiotic |
| Streptomycin | *Streptomyces griseus* | antibiotic |
| Cephalosporin | *Cephalosporium acrimonium* | antibiotic |
| Griseofulvin | *Penicillium griseofulvin* | antifungal antibiotic |
| Cyclospori A | *Tolypocladium inflatum* | immunosuppressant |
| Gibberellin | *Gibberella fujikuroi* | plant growth regulator |

Secondary metabolism may be repressed in certain cases. Glucose represses the production of actinomycin, penicillin, neomycin and streptomycin; phosphate represses streptomycin and tetracyclin production.

Hence, the culture medium for secondary metabolite production should be carefully chosen.

### Food Industry Products

A very wide range of innumerable products of the food industry, such as sour cream, yoghurt, cheeses, fermented meats, bread and other bakery products, alcoholic beverages, vinegar, fermented vegetables and pickles, etc., are produced through microbial fermentation processes. The efficiency of the strains of the organisms used, and the processes are being continuously improved to market quality products at more reasonable costs.

### Production Enzymes

Industrial production of enzymes is needed for the commercial production of food and beverages. Enzymes are also used in clinical or industrial analysis and now they are even added to washing powders (cellulase, protease, lipase).

Enzymes may be produced by microbial, plant or animal cultures. Even plant and animal enzymes can be produced by microbial fermentation.

While most enzymes are produced in the tropophase, some like the amylases (by *Bacillus stearothermophilus*) are produced in the idiophase, and hence are secondary metabolites.

***Table:*** *Examples of commercially produced enzymes*

| Organism | Enzyme |
|---|---|
| *Aspergillus oryzae* | Amylases |
| *Aspergillus niger* | Glucamylase |
| *Trichoderma reesii* | Cellulase |
| *Saccharomyces cerevisiea* | Invertase |
| *Kluyveromyces fragilis* | Lactase |
| *Saccharomycopsis lipolytica* | Lipase |
| *Aspergillus* species | Pectinases and proteases |
| *Bacillus* species | Proteases |
| *Mucor pusillus* | Microbial rennet |
| *Mucor meihei* | Microbial rennet |

### Recombinant Products

Recombinant DNA technology has made it possible to introduce genes from any organism into micro-organisms and *vice versa*, resulting in transgenic organisms and the latter are made to produce the gene product. Genetically manipulated *Escherichia coli, Saccharomyces cerevisiae*, other yeasts and even filamentous fungi are now being used to produce interferon, insulin, human serum albumin, and several other products.

### *Biotransformation*

Production of a structurally similar compound from a particular one, during the fermentation process is transformation, or biotransformation, or bioconversion. The oldest instance of this process is the production of acetic acid from ethanol. Immobilised plant cells may be used for biotransformation. Using alginate as the immobilising polymer, digitoxin from *Digitalis lanata* was converted into digoxin, which is a therapeutic agent in great demand. Similarly, codeinone was converted into codeine and tyrosine from *Mucuna pruriens* was converted into DOPA.

### *Elicitors*

It is possible to induce production or enhance production of a compound in cultures by using elicitors, which may be micro-organisms. For example, *Saccharomyces cerevisiae* was an efficient elicitor in the production of glyceollin (*Glycine max*) and berberine (*Thalictrum rugosum*). *Rhizopus arrhizus* trebled diosgenin production by *Dioscorea deltoidea*. The production of morphine and codeine by *Papaver somniferum* was increased 18 times by *Verticillium dahliae*.

## Genetic Improvement of Fermentation Processes

The genome of the organism ultimately controls its metabolism. Although improved fermenter engineering design and optimal cultural conditions can quantitatively enhance the microbial products, this will only be up to a limit. Genetic improvement of the organism is fundamental to the success of fermentation technology. Mutation and recombination are the two ways to meet this end.

### *Mutation*

A certain amount of mutational change in the genome occurs as a natural process, though the probability is small. Exposing a culture of a micro-organism to UV light, ionising radiation or certain chemicals, enhances the rate of occurrence of mutations. But it is a tremendous task for the industrial geneticist to screen the very large number of randomly produced mutants and to select the ones with the desired qualities. The synthesis of a number of products of cell metabolism is controlled by a 'feed-back inhibition'. When a compound reaches a particular level of accumulation, its synthesis is stopped. Synthesis starts again when the level of the compound falls below the specific level. If a mutant is produced, in which the feedback signalling is suppressed, the product is synthesised continuously. By such a manipulation, a high producing strain of *Corynebacterium glutamacium* was developed to recover very high quantities of lysine. Such strains that do not produce controlling end products are called auxotrophs.

### Recombination

Recombination is defined as any process that brings together genes from different sources.

A strain of *Brevibacterium flavum* is a high producer of lysine, but is limited by its poor capacity to absorb glucose. Another strain of the bacterium, which is an efficient absorber of glucose but which does not produce lysine, was used to develop a recombinant strain, through protoplast fusion. The new strain utilises high levels of glucose and yields higher levels of lysine. A gene for the synthesis of phenylalanine was transferred to a chosen strain of *Escherichia coli,* which was a non-producer, but a good experimental and production tool.

Transformation of a high cephalosporin producing strain of *Cephalosporium acremonium* with a plasmid containing the gene REXH has significantly increased the titre.

A number of human proteins, such as insulin, human growth hormone, bone growth factor, alpha, beta and gamma interferons, interleukin-2, tumour necrosis factor, tissue plasminogen activator, blood clotting factor VIII, epidermal growth factor, granulocyte colony stimulating factor, erythropoietin, etc., are being produced through recombinant micro-organisms.

### DNA Manipulation

*In vitro* DNA technology was used to increase the number of copies of a critical pathway gene (operon), as for example the production of threonine in *Escherichia coli*, at rates 40 to 50 times higher than usual.

Fermentation technology is a very vibrant and fast growing area of biotechnology, absorbing an ever increasing processes and products. With a longer history than any area of biological sciences, fermentation technology has a longer and brighter future, in the service of mankind, covering such important areas as food and medicine.

## The Fly Ash on Indian Biotechnology from the European Commission's Policy on Genetically Engineered Products

The recent passing of a new legislation by the European Commission lifting the *de facto* moratorium on the commercialization genetically engineered organisms (GEOs, GMOs) in Europe, resulted in a sigh of relief, that was stifled midway, on account of the largely impractical regulations associated with it. The Ministers for Agriculture of the member countries have endorsed the new regulations. Most of these regulations relate to segregation and labeling of GE products and have not much to do with science or biosafety.

The EC's *de facto* moratorium of the past six years prevented commercial releases of GEOs in Europe, and hampered research on them. Both private investment and expertise largely moved over to the US. Nevertheless, over 200 traits were identified for transgenic technology in various crops in the EU countries. By the end of June 2003 the EC received 64 reports of studies on biosecurity and yield performance of transgenic crops, ready for commercialization, from the member countries.

Belying the hope for a rational climate from the EC, the new policy turned out to be a pyrrhic victory to the pro-tech lobby. What is more perplexing is that the stringent regulations were framed ignoring overwhelming evidence, much of which originated from Europe itself, on the safety of GE crops. In 2001, the EC released results of a 15-year study, costing US $ 64 million, and involving more than 400 research teams and 81 projects. This report concluded that GE products pose no more risk to human health or the environment than conventional crops. The Strategy Unit of Cabinet Office of UK has also released a new report on the costs and benefits of GE crops. In addition, the recent GM-Nation report involving 16,000 interviews, in 15 cities and in 11 languages, is considered as the most balanced set of recommendations and conclusions to go forward with the commercialization of GE crops. So far, extensive and intensive research in the US, Australia and elsewhere, on the probable risks of GE technology, has not brought out any adverse effects and none of the fears expressed by anti-tech activists were proved even marginally. Such overwhelming evidence should have been sufficient to soften up on GEO regulations by the EC.

The silver lining is the EC's stand that public authorities cannot ban farmers from planting genetically modified crops. The EC also published guidelines for the development of strategies and practices to ensure the co-existence of GE crops with conventional and organic farming. This will provide some support to those farmers in the EU, who want to embrace the technology, and takes some wind out of the sails on anti-tech activists in other countries.

Biosecurity regulations should be objective to the extent of ensuring biosafety and environmental safety. A strict compliance of must be enforced. But the regulations must themselves be based in science, risk specific, rational and practical. While the scientific community is reasonably clear and reassured on biosecurity issues, the objective of the whole regulatory process is to reassure the public on the safety of GE products. This objective largely remains unachieved, without appropriate public awareness and education programmes about the purpose, benefits and risk mitigation related to GEOs. Public education is the critical need of the hour to save the public from being misguided by the negative propaganda of vested interest groups.

Too rigid and impractical regulations will result in either no product getting into commercialization thus denying the potential benefits to the public or the regulations becoming lackadaisical. It would be akin to the statutory warning on cigarette packets, which every smoker sees but ignores it. If a product is bad in the public interest it should be banned altogether. Such a step requires a strong conviction, courage and political will, which in the context of GEOs, is totally absent. Merely making noises and creating scenes, when one does not have the responsibility of being answerable for actions, is quite a convenient situation, for the rabble-rousers.

The new regulations of the EC will certainly put the member countries at a lot of disadvantage in the matter of GEO development and deployment. This will also result in infringement of certain regulations of the WTO. These consequences would not be confined to the EU countries, but have a telling effects on the Asian and African countries, with major export interest in the agriculture sector. Even now, several member countries of the EU have not fully complied with their responsibilities. The EC has recently decided to take eleven countries to the European Court of Justice, for failing to adopt and notify national legislation implementing the EU law on the deliberate release of genetically modified organisms (GMOs) into the environment. This is in addition to several other infringements by member countries. Having spent enormous amounts of money and time on developing and testing several transgenic products, the defaulting member countries are naturally wary of the effects of the new stringent rules they have to pass and implement under the new guidelines of the EC.

Spain would suffer most from EC's rigid regulations, as 34 out of the 64 European projects, came from Spain. Of these 34 reports, 18 are on rice (15 on the yield increase and three on abiotic stress), 14 on maize and one on gene stacked (Cry 1F/Cry 1Ac) cotton. France has 16 GEOs, Germany six, while others have one to three transgenics, ready for commercial release, once (and if) the green signal comes from the

EC. Left to the respective countries, the stringency of the biosecurity regulations each country makes, would depend much on the economic stakes at risk for that particular country.

Infringement of safety regulations does not seem to be confined to just any one country or continent. A fifth of the Bt maize farmers appear to have flouted federal regulation, by planting refuge at either below-regulation levels (19 per cent) or much worse even without any refuge at all (13 per cent), in the Midwest US. Several Indian farmers, who cultivated Mahyco/Monsanto's Bollgard Bt-cotton, planted less than regulatory requirement of refuge and in some cases no refuge at all. The obvious reason is there for all to see. If one makes impractical and scientifically untenable regulations, they will be defied or not complied with in its entirety. Market forces have their own strange ways of breaking them or circumventing them. This has happened time and time again in all spheres of economic activity, and agriculture is no exception.

Biotechnology in India and other developing countries has suffered serious damage from the past policy of the EC. It is not that EU dictates policy on GEOs to the developing countries, but its policy has repercussions on agricultural trade and development. Anti-technologists and the Greens take advantage of EC's stringency. They have already twisted EC's *de facto* moratorium into a virtual ban. The refusal of USAID by Zimbabwe, a fall out of EC's policy, was repeated in India. When the public, and even the regulatory bodies, repeatedly hear that "Europe has banned GEOs", naturally doubts and fear rule the roost.

Vested interest from different quarters used this position to advantage. We cannot derive much happiness from the recent lifting of the *de facto* moratorium by the EC, since the anti-tech activists would now shift their position into demanding for regulatory procedures including labeling, as rigid or even stricter than those imposed by the EC on its member countries.

This will have serious consequences for the developing countries that have no infrastructure, expertise and financial resources, to comply with such regulations. In fact, several activists in Indian have demanded an absurd and scientifically unwarranted regulatory regimen, including an outright ban on GEOs. Such postures had an unfortunate effect on the functioning of the Genetic Engineering Approval Committee (GEAC), the highest regulatory authority in India. The GEAC has used the Precautionary Principle more often than most regulatory bodies, and has been making inexplicable and irrational decisions on the release of GE crops in India.

India can certainly find a sensible position between the rigid policy of the EC and the rather lax policy of China or some of the African countries. India has the means to handle GEOs in a rational manner, provided its scientific community, rather than its bureaucracy, frames and implements policy. This is possible only if there was political will to prevent inter-continental and multi-national corporate politics, which mainly operate in the garb of anti-tech postures, from interfering with decision-making. Decisions should be timely, rational, consistent and transparent. Regulations should be based in science and made in relevance to India's crops and conditions. All stakeholders must be a party to the process of making policy and regulations. This is the only way to save Indian agriculture and give the farmers the benefits of a promising technology.

### Functional Diversity

While dealing with microorganisms it is appropriate to give equal emphasis to structural (morphological) and functional diversity. Based on their activities, the fungi can be grouped as cellulase producers, lignin degraders, nitrate reducers, pectin degraders, P-solubilizers, iron chelaters, Penicillin producers &c. The common traits between a set of fungi implies the presence of certain common gene clusters which have value in a particular ecological niche (Watts *et. al.*, 1999).

### Habitat Diversity

The mycoflora of unique ecological niches have some common features and it will be very rewarding to explore fungal diversity in habitats such as thermophilic environments, e.g., hot springs, thermal vents, sun-heated soils, compost pits, self-heated coal refuse piles, steam line discharge sites, etc. The thermophilic fungi and bacteria grow at temperatures between 40-60 C are the sources of thermostable enzymes. Other ecological niches to be explored are western ghats, the Himalayan ranges, marine ecosystems, mangroves, coral reefs, sand dunes, industrial effluent contaminated soils, ant hills, refineries, activated sludge, insects and several other natural sources. Endophytes of plants and animals are very meagerly understood.

### Bioprospecting

The process of looking for new forms of life and/or their products is called Bioprospecting. The techniques of fungal exploration have remained by and large conventional. The methods have mainly included the use of different recipes for culture media along with techniques such as dilution plating, soil plating, leaf shred inoculation &c. Direct incubation and baiting techniques have also been in use. The emphasis is now being laid on enrichment techniques wherein the medium used mimics the natural habitat in nutritional status. One can isolate nitrate reducers using a medium enriched with nitrate and minimal carbon source. Reduction of carbon source is also a means of keeping the contaminants away. Addition of antibiotics in enrichment cultures is another way of keeping away the weedy fungi and bacteria. Enrichment cultures have often been used to isolate fungi capable of degrading certain specific xenobiotics.

The xenobiotic in question may be used as the sole carbon source in the medium so that only the fungi with degradative capacities will grow. In liquid culture, competition for the particular substrate will lead to the enrichment of the microbial strain that is able to grow faster than others. One flaw in the enrichment technique is that it leads often to growth of bacteria even when fungi with good potential are present. It is because of the capacity of bacteria to grow faster than fungi. Besides, some fungi may not utilize the degraded product as the sole source of carbon or as a supplementary source of carbon, as in the case of the white rot fungi (Jefries *et al.*, 1981). It may be therefore necessary to use bacteriostatic compounds and supplementary carbon sources to isolate xenobiotic degrading fungi from a natural source.

When it is not possible to grow a fungus by any means, isolation of the gene pool from the environment may be very rewarding. Environmental DNA is now being harnessed for commercial production of certain substances without the need to grow the actual microorganism. The gene of interest separated from environmental DNA can be tagged onto a vector and cloned into a suitable expression host to yield the product. More work has been done this area with bacteria than fungi.

Bioprospecting is important because fungi and other microorganisms are the largest future source for various products and processes. They are of critical importance to the sustainability of life in nature because they are involved in the recycling of elements and in producing and consuming gases such as oxygen and carbon dioxide so important in maintaining our climate. Fungi have been employed for the industrial production of compounds of intrinsic value such as antibiotics, enzymes, hormones, immunosuppressants, food additives and colorants. They have also been employed as biopesticides and biofertilizers. They have been the sources of microbial protein and oil and animal feed. Edible fungi have been known as long as human civilization. The exploration of the extent of fungal diversity on planet earth is indeed a stupendous, but surely a very rewarding task.

### Bioremediation

Bioremediation is a pollution control technology that uses biological systems to catalize the degradation or transformation of various toxic chemicals to less harmful forms. The general approaches to bioremediation are to enhance natural biodegradation by native organisms (intrinsic bioremediation), to carry out environmental modification by applying nutrients or aeration (biostimulation) or through addition of microorganisms (bioaugmentation). Unlike conventional technologies, bioremediation can be carried out on-site. Bioremediation is limited in the number of toxic materials it can handle (Hart, 1996), but where applicable, it is cost-effective (Atlas & Unterman, 19977).

Biodegradation, mineralization, bioremediation, biodeterioration, biotransformation, bioaccumulation and biosorption are some terms with minor differences but often overlappingly used. Biodegradation is the general term used for all biologically mediated break down of chemical compounds and complete biodegradation leads to mineralization. Biotransformation is a step in the biochemical pathway that leads to the conversion of a molecule (precursor) into a product. A series of such steps are required for a biochemical pathway. In environmental terms, it is importance whether the product is less harmful or not. Biodeterioration refers usually to the breakdown of economically useful compounds but often the term has been used to refer to the degradation of normally resistant substances such as metals, plastics, drugs, cosmetics, painting, sculpture, wood products and equipment (Rose, 1981). Bioremediation refers to the use of biological systems to degrade toxic compounds in the environment. Bioaccumulation or biosorption is the accumulation of the toxic compounds inside the cell without any degradation of the toxic molecule. This method can be effective in aquatic environments where the organisms can be removed after being loaded with the toxic substance.

The fungi are unique among microorganisms in that they secrete a variety of extracellular enzymes. The decomposition of lignocellulose is rated as the most important degradative event in the carbon cycle of earth. Enormous literature exists on the role of fungi in the carbon and nitrogen cycles of nature. The role of fungi in the degradation of complex carbon compounds such as starch, cellulose, pectin, lignin, lignocellulose, inulin, xylan, araban etc., is well known. *Trichoderma reesei* is known to possess the complete set of enzymes required to breakdown cellulose to glucose. Degradation of lignocellulose is the characteristic of several basidiomycetous fungi.

### Fungi in Bioremediation

Fungi are good in the accumulation of heavy metals such as cadmium, copper, mercury, lead and zinc. Systems using *Rhizopus arrhizus* have been developed for treating uranium and thorium.

The ability of fungi to transform a wide variety of hazardous chemicals has aroused interest in using them in bioremediation (Alexander, 1994). The white rot fungi are unique among eukaryotes for having evolved nonspecific methods for the degradation of lignin; curiously they do not use lignin as a carbon source for their growth (Kirk *et al.*, 1976). Lignin degradation is, therefore, essentially a secondary metabolic process, not required for the main growth process. Lamar *et al.* (1993) compared the abilities of three lignin-degrading fungi, *Phanerochaete chrysosporium, P. sordida* and *Tramates hirsuta* to degrade PCP (Pentachlorophenyl) and creosote in soil. Inoculation of soil with 10% (wt/wt) *Phanerochaete sordida* resulted in the greatest decrease of PCP and creosote. *P. sordida* was also most useful in the degradation of PAHs (Polycyclic aromatic hydrocarbons) from soil. Davis *et al.,* (1993) showed that *P. sordida* was capable of degrading efficiently the three ring PAHs, but less efficiently the four-ring PAHs.

*Phanerochaete chrysosporium* has been shown to degrade a number of toxic xenobiotics such as aromatic hydrocarbons (Benzo alpha pyrene, Phenanthrene, Pyrene) chlorinated organics (Alkyl halide insecticides, Chloroanilines, DDT, Pentachlorophenols, Trichlorophenol, Polychlorinated biphenyls, Trichlorophenoxyacetic acid), nitrogen aromatics (2,4-Dinitrotoluene, 2,4,6-Trinitrotoluene-TNT) and several miscellaneous compounds such as sulfonated azodyes. Several enzymes that are released such as laccases, polyphenol oxidases, lignin peroxidases etc. play a role in the degradative process. In addition, a variety of intracellular enzymes such as reductases, methyl transferases and cytochrome oxygenases are known to play a role in xenobiotic degradation (Barr & Aust, 1994).

*Phanerochaete chrysosporium* has been shown to effect the bioleaching of organic dyes (Nigam *et al.,*1995). Pauli ollikka *et al.* (1993) have also shown the decolorization of azo-triphenyl methane dyes by lignin peroxidase produced by *P. chrysosporium.* Sami and Radhaune (1995) have demonstrated the role of lignin peroxidase and manganese peroxidase from *P. chrysosporium* in the decolorization of olive mill wastewater. The work carried out in our laboratory has shown that *Phanerochaete chrysosporium* and microbial consortia were effective in colour removal from textile dye effluents The fungus caused 80% decolorization in broth containing 2.5% of effluent. There was reduction in BOD and COD values. A local isolate of *Fusarium sp.*, caused various degrees of decolorization ranging from 35 to 85 %.

Among the fungal systems, *Phanerochaete chrysosporium* is emerging as the model system for bioremediation. The basidiomycetous fungus *Pleurotus ostreatus* has been shown to produce an extracellular hydrogen peroxide dependent lignolytic enzyme that removes the colour due to remozol brilliant blue. Oxidative enzymes play a very major role in biodegradation. Other fungi that can be used in bioremediation are obviously the members of Zygomycetes e.g., the mucoraceous fungi and the arbuscular mycorrhizal fungi. Aquatic fungi and anaerobic fungi are the other candidates for bioremediation.

Among other fungi used in bioremediation, the yeasts, e.g., *Candida tropicalis, Saccharomyces cerevisiae, S. carlbergensis* and *Candida utilis* are important in clearing industrial effluents of unwanted chemicals. *Agaricus bisporus* and *Lentinus oloides* are important in lignocellulose decomposition. *Corius versicolor* is important in cleaning up pulp and paper mill wastes. Consortia of fungi and bacteria (usually uncharacterised) are used in composting, the most useful waste disposal practice. Phenolic azo dyes have been shown to be oxidized by the enzyme laccase produced by *Pyricularia oryzae*.

## Future Outlook

Whenever bioremediation figures as the topic of discussion, bacterial agents come into focus and fungi are much less studied. One should realize, however, the greater potential of fungi by virtue of their aggressive growth, greater biomass production and extensive hyphal reach in soil. More research will be focused in future on using the diverse fungal flora for bioremediation. The work on the microbial diversity in the electroplating industrial effluents is going on in our laboratory. There are several promising fungi that can degrade zinc, cyanide and chromium in these effluents.

Future work will be more focussed on the biotechnological aspects. It may be possible to clone the highly efficient degradative enzyme producing genes into bacteria and conversely, bacterial genes can be transferred to fungi which are suitable. The high surface-to-cell ratio of filamentous fungi makes them better degraders under certain niches like contaminated soils. Fungi have been shown to even solubilize partially coal, a highly polymeric substance more complex than lignin. There is no doubt, therefore, regarding fungi being harnessed more and more in environmental bioremediation work in future.

## India's Biotech Aspirations

***Cover Story:*** India looks to the "BT" of biotech to lead it into the post-IT economy. But significant barriers must be overcome to achieve this potential. The alignment of a vast pool of scientific talent, a worldclass information technology industry, and a vibrant generic pharmaceutical sector are positioning India to emerge as a significant spot on the global biotech map.

India is home to the largest number of English speaking scientists outside the U.S. and in the U.S. Indians are well represented on the boards of startup high tech companies and in academic research laboratories. India's success in IT- the south Asian country grew its computer software and services industry from zero to more than $7 billion in exports over the last decade-may lend credibility to its biotech aspirations, as do the strong business and cultural connections between the subcontinent and Silicon Valley. But building biotech companies is far more complex. And to achieve its potential in biotech, India will have to overcome some significant barriers, including a confused regulatory environment, uncertainties about intellectual property protection and the slow pace of integration between academic and commercial science.

***Hitting the radar:*** The involvement of Indian companies and researchers in two of the year's biggest biotech public policy stories, bioterrorism and stem cells, provides signs that the gaps between university lab and commercial markets are being bridged. It also illustrates the range of biomedical activity underway in India.

When U.S. Sen. Charles Schumer (DN. Y.) began his campaign to increase supplies of Bayer AG's Cipro ciprofloxacin antibiotic, his staff contacted RanbaxyLaboratories Ltd. in New Delhi, one of several Indian companies that have received tentative FDA approval to market generic ciprofloxacin as soon as Bayer's patent expires. In addition, researchers at the Centre for Biotechnology at Jawaharlal Nehru University in New Delhi announced in late October that they have been working for four years on a recombinant anthrax vaccine, have completed preclinical studies and soon would start Phase I trials in collaboration with Panacea Biotech Ltd. (New Delhi).

In the stem cell arena, Reliance Life Sciences (Mumbai), a recently formed division of one of India's largest industrial conglomerates, and the National Centre for Biological Sciences (Bangalore) are among 11 institutions in five countries named by NIH as sources of embryonic stem cells that can be used in U.S. government funded research. Business models Unlike the U.S. and Europe, where biotech companies are typically started by academic scientists, Indian companies have taken four distinct routes to the biotech space: diversifying from large scale industrial activities such as chemical manufacturing; creating startups focused on indigenous production of first generation recombinant proteins; investing profits and expertise from generic pharmaceutical manufacturing to move into discovery; and grafting biology branches onto strong IT stems to create bioinformatics and genomics companies.

Biocon Ltd., the first and largest biotech company in India, started in 1978 formulating industrial enzymes provided by an Irish company. The company's founder and CEO, Kiran MazumdarShaw, has taken it through a complex pathway that included becoming a joint venture with Unilever plc, acquisition of Unilver's stake by ICI plc, and emerging in 1998 as an independent company.

Using a proprietary solid surface fermentation technology, Biocon retains its roots in the world markets for food and industrial enzymes. It also has turned its fermentation technology to the manufacture of drugs, developing a pipeline of statins and winning U.S. FDA approval this year to market generic lovastatin for cholesterol reduction. Biocon also has created contract drug discovery, clinical trials, genomics and chemistry research units and sister companies. Biocon's Syngene unit has 150 scientists working on drug discovery, chemical synthesis, molecular biology, gene expression and related areas. Indian companies entered the recombinant protein field in response to urgent needs for inexpensive vaccines to treat easily prevented infectious diseases that plague the nation. Varaprasad Reddy, an electronic engineer with no experience in the life sciences and modest financial resources, initiated a project to produce a recombinant hepatitis B vaccine in 1992, sending six researchers to the labs of Indian molecular biologists in the U.S. for training.

Hyderabad, Bharat Biotech International Ltd., and Biological E Ltd., acquired competency in recombinant protein R&D and created their own manufacturing technologies. In addition to these startups, India's fifth largest pharmaceuticals company, Wockhardt Ltd. (Mumbai), manufactures and sells a hepatitis B vaccine based on licenses from Rhein Biotech NV (NMarkt:RBO, Maastricht, the Netherlands). The Indian vaccine startups now are turning to indigenous versions of other recombinant proteins, including granulocyte colony stimulating factor (GCSF) and other growth factors, erythropoietin (EPO) and insulin. Dr. Reddy's Laborato In 1995 he secured lab space and technical assistance from the Centre for Cellular and Molecular Biology (CCMB, Hyderabad) and funding from the Bank of Oman for construction of a manufacturing plant. Two years later, India's first genetically engineered vaccine was on the market, selling for 20 times less than the imported product. Today, Shantha Biotechnics Pvt. Ltd. has 360 employees, manufactures interferon IIa and has five other proteins in the pipeline. Pfizer Inc. distributes Shantha's hepatitis B vaccine in India and has right of first refusal on the company's new products.

In the process of developing low priced hepatitis B vaccines to compete with imported products that most Indians could not afford, Shantha and its rivals in Indian research centres Traditional barriers between government funded research laboratories and industry are crumbling, enabling Indian biotech companies to tap into expertise, resources and manpower. The Council of Scientific and Industrial Research (CSIR) operates a network of government laboratories with mandates to collaborate with industry. CSIR's Centre for Cellular & Molecular Biology incubated India's first recombinant protein product, Shantha Biotechnics' hepatits B vaccine, and it has numerous industrial relationships, including a joint venture with Biological E and Amersham Pharmacia to build DNA microarrays. The National Brain Research Centre, operated by the Department of Biotechnology, part of the national government, conducts intramural and supports extramural neuroscience research.

Many of the country's major universities, including Jawaharlal Nehru University (New Delhi) and the University of Hyderabad, have strong molecular biology and chemistry expertise. The Indian Institute of Science (Bangalore), a private institution, operates a Centre for Scientific and Industrial Consultancy to facilitate technology transfer and industrysponsored research.

Shantha and Bharat are positioning themselves to partner with U.S. generic companies to enter the American and European markets when patents on these products expire. They also are collaborating with Indian and western companies and research funders to develop new prophylactic and therapeutic vaccines for AIDS, and for malaria, rotavirus and other diseases that are relevant for tropical developing countries. While the first indigenously produced recombinant biologic hit the market in 1997, Indian generic companies have deep roots- and much deeper pockets - than the startups. Protected by high tariffs and a Sovietstyle command economy, featuring five year plans targeting capital investments to government affiliated businesses and a rigid system of licenses and permits controlling commerce, India built a large chemical manufacturing industry in the 1960s and 1970s.

The chemical sector turned its attention to bulk drug components in the 1980s when the Indian economic system crumbled in synch with the Soviet Union's, and in 1990 the government initiated reforms to reduce restrictions on industrial development, increase competition and open domestic markets to imports. Stimulated by the new competitive environment and helped by the lack of product patent protection for pharmaceuticals, Indian industry turned its chemical manufacturing knowhow to generic drugs. Perhaps paradoxically, the generic companies became adept at devising novel, efficient drug manufacturing techniques in part because process patents are valid in India. Today, the larger generic companies are moving into biologics and have discovery programs. RDY, which was the first company to get FDA approval for generic fluoxetine (Prozac), has discovery efforts targeted at diabetes, inflammation, lipid metabolism, oncology, and cardiovascular disease. The company plans to launch monoclonal antibody products in India by 2004, according to Chairman Anji Reddy. In addition, RDY has licensed two insulin sensitizers to Novo Nordisk A/S, which is slated to start Phase III trials in North America in the first quarter of 2002. RDY also has established a research subsidiary in Atlanta, Reddy US Therapeutics Inc., as well as a contract research subsidiary that will focus on genomics. Last week, Wockhardt announced that it filed an IND application with the Drug Controller General of India to conduct clinical trials of WCK771, a broadspectrum antibacterial agent with activity against both Gram positive and Gram negative pathogens. The company is developing two other antibacterial agents. Ranbaxy, India's largest pharma company with 2000 sales of more than $500 million, also views innovation as key to its future. The company has branched out from creating new formulations of existing drugs and has half a dozen molecules under development, Brian Tempest, president pharmaceuticals, told BioCentury.

Ranbaxy has collaborations with several U.S. and European companies to develop new formulations and delivery technologies. For example, Ranbaxy and Vectura Ltd. (Bath, U.K.) announced in June that the Indian company's Ranbaxy B.V. subsidiary (Antilles, the Netherlands) will develop oral formulations using Vectura's controlledrelease drug delivery technology, with Ranbaxy providing clinical development, scaleup, manufacturing and marketing expertise. Partnering model Ranbaxy is looking for global partners to help it move more aggressively into the biotech space and can bring costeffective basic science, clinical research and manufacturing capabilities to the table, Tempest said.

"From the U.S., it looks like cost of manufacture is our edge. But our cutting edge is actually the low cost of innovation," he said. "A number of companies in the U.S. have banks of molecules, including some that are not commercially viable to exploit. If they partner with a big pharma company, they won't get anything. On the other hand, an Indian company can bring a huge ability to manufacture and to conduct clinical trials." Guljit Chaudhri, senior vice president of strategic business development at a basic and specialty chemicals manufacturer that is diversifying into pharmaceuticals, Jubilant Organosys Ltd. (New Delhi), also sees a mesh between the capabilities and needs of Indian companies and global biopharmaceuticals players.

"Internationally, the cost of R&D is constantly increasing and the pace of new discoveries is slowing down," she noted. "This provides an opportunity for Indian players to lever there is a fairly substantial opportunity for the Indian firms and for U.S. firms because they can go further down the food chain. India has a large untapped pool of talent in protein chemistry, medicinal chemistry and classic synthetic and organic chemistry." Some U.S. and European collaborations are tapping into Indian expertise, some reflecting the expatriot links already seen between India and Silicon Valley. For example, the CCMB recently won a contract from Onconova Therapeutics Inc. (Princeton, N.J.) to create transgenic fruit fly high throughput assay systems and use them to screen drug targets for anticancer effects.

Onconova president Ramesh Kumar said he was attracted by CCMB's capabilities, low cost, and eagerness. "We could collaborate with companies or academic groups in the U.S., but we found that most of the good academic labs were working with target validation companies and those companies were working with large pharma companies," he said. According to Kumar, CCMB is providing worldclass science at a cost 50fold lower than Onconova would have to pay in the U.S.

The company also raised much of its startup funding in India, including $3 million from Zydus Cadila Healthcare Ltd. (Ahmedabad), a generic company that is moving into discovery and biotech. Zydus also is partnering in Europe, in January announcing a deal to develop an antibiotic with Pantheco A/S (Copenhagen, Denmark). Zydus will undertake chemistry, preliminary screening and initial characterization of compounds with antibacterial activity. Pantheco will perform preclinical and early clinical development. Each company will cover its own R&D costs, and profits will be shared 5050. In 1998, when Symyx Technologies Inc. (SMMX, Santa Clara, Calif.) considered out sourcing the synthesis of organic chemicals used in its discovery programs, it evaluated companies in the U.S. and offshore.

"We did a parallel evaluation and found that the folks at Biocon's Syngene were certainly as fast as folks elsewhere, had a higher success rate with the compounds they attempted and were almost an order of magnitude less expensive," said Peter Cohan, vice president for discovery. Cohan told BioCentury that SMMX did not sacrifice quality or speed. "A lot of people believe that work done in China, manage their lowcost advantage and strong chemical synthesis skills to enter into collaborative research projects with multinational companies." Ajay Dhankar, an industry analyst for McKinsey and Co. in Mumbai, agreed that the ability to provide lowcost research, high quality chemistry, and informatics and manufacturing expertise could make Indian companies attractive partners for mid sized U.S. and European biotechs that are looking for help climbing the value chain.

"If you take a medium sized biotech company in the U.S. that is reasonably cash rich but doesn't have money to throw away, maybe it has five targets and 20 leads. It can't develop them all, but it makes no sense monetarily to license them out to big pharma that early. They could be in a good position to form an alliance with an Indian company that could develop their leads at a much lower cost," he said.

The Indian software industry was built on a service model, initially "body shopping" programmers to the U.S. and Europe, and then doing contract programming on legacy computers and providing Y2K solutions. Dhankar argued that the service model similarly could jump start the Indian biotech sector. "India has created a very successful model around remote IT services. We see an analog in biotech happening where we expect that Indian companies will be able to provide high value services to global pharma and biotech companies," he said.

"Most biotech firms in the U.S., particularly those focused on small molecules, lack chemistry skills, the downstream capabilities crucial for developing products. They are starting to realize that they've been focused on upstream activities and haven't been focused on lead optimization," Dhankar added. "We expect that if a certain number of Indian companies can focus on this as an area of development, India or elsewhere is second rate. The fact of the matter is there are highly talented, motivated scientists offshore who are looking for opportunities to succeed and delight their customers and Syngene is one of those companies," he said.

Other Sygene clients include Abbott Laboratories (ABT, Abott Park, Ill.), Neogenesis Drug Discovery Inc. (Cambridge, Mass.), AstraZeneca plc (AZN, London, U.K.), GlaxoSmithKline plc (GSK, London, U.K), Affymax (Palo Alto, Calif.), BristolMyers Squibb Co.

(BMY, Princeton, N.J.), Dow Biotech (San Diego, Calif.) and others. Following on this line, in September, RDY chairman Reddy announced the formation of a new company, Aurigene Discovery Technologies, to provide proteomics and discovery services to biopharmaceutical companies.

The informatics crossover Given India's strong IT base, bioinformatics is an obvious choice for entrepreneurs seeking a ticket into the biotech space. Scores of large IT companies have established bioinformatics units and Bangalore is bristling with bioinformatics startups. "There are opportunities in India for data mining, gene annotation, and the development of software interfaces. These will require enormous computing power and money and there is no better place than India to do this," said D.P. Verma, a prominent plant molecular biologist. While a great deal of the action may prove to be froth, India is building substantial infrastructure and some companies are linking up with the biomedical expertise necessary to make real businesses. The Department of Biotechnology plans to award a $120 million contract for a supercomputer based network to link 11 bioinformatics centres and provide researchers with access to genomics and proteomics databases. Satyam Computer Services Ltd. (SAY, Hyderabad) is working with the CCMB to develop annotation and other genomics data products and to offer bioinformatics services to international clients.

There are hundreds of bioinformatics companies springing up like mushrooms, but most don't have any understanding of what customers need. They think they can just put a bunch of programmers and cally retarded India's progress in biotechnology, according to Pushpa Bhargava, a molecular biologist who founded the CCMB. In 1988, Bhargava said, he estimated that low wages and overhead would allow India to sequence the entire human genome for a small fraction of the cost in industrialized countries. However, he said, government officials rejected his plans and other proposals to take bold scientific initiatives.

Exploiting the possibilities opened by the genome sequence provides India with a second chance, particularly in areas like proteomics that rely on IT expertise. And in a departure from precedent, the challenge is being grasped by both the public and private sectors, according to Bhargava, who advises several biotech startups. While advances in technology have reduced the relevance of lowcost manpower, India still can apply a different kind of human resources to genome discovery. "We have the largest human biodiversity in the world," Bhargava said. In this subcontinent there are close to 600 welldefined ethnic groups which over the centuries, for millennia, have kept their identity. Twenty percent of India is tribal and they have maintained their identity. That's marvelous in terms of material, something no other country can match," he said.

Thus Biocon is working with Surromed Inc. (Palo Alto, Calif.) to identify biomarkers for diabetes in the Indian population. Nicholas Piramal India Ltd., a Mumbai pharmaceutical company, has formed a partnership with the government's Centre of Biotechnology to conduct genomic research with the nation's diverse populations and to explore India's traditional medicines.

Barriers to overcome Viewed from a distance, the impressive technical achievements of the Indian elite can obscure the difficulties of conducting 21st century science and business in a developing country. While it has created nuclear weapons and written much of the software that coordinates the financial transactions of multinational companies, India spends $3 per capita annually on drugs. The future of the Indian biotech industry depends to a great extent on the government's ability to balance competing demands for cheap drugs with the imperative to stimulate innovation by enforcing intellectual property rights. computers in a room and start doing bioinformatics," said V.N. Balaji, CEO of the Jubilant Biosys Ltd. subsidiary (Bangalore) of Jubilant Organosys.

He predicted that while most of these companies will disappear, real opportunities will remain. "Most of the U.S. and European genomics and bioinformatics companies are turning themselves into pharmaceutical companies and at the same time the decreasing costs of microarray chips and moves toward in silico research will generate enormous amounts of data," Balaji said. Jubilant's model is based on performing contract bioinformatics services, at client sites and in Bangalore, creating and curating databases, building customized tools, and performing contract research. Sreenivas Devidas, vice president of business development and strategy at Strand Genomics Pvt. Ltd. (Bangalore), predicted that the market will break into two segments, with some companies providing lowmargin, lowend services such as annotation, and a handful of companies providing highend services.

Devidas, a former vice president and life sciences advisor to venture fund ConnectCapital Holdings (Mumbai), said Strand is positioning itself at the top end. Strand, a spinoff from the Indian Institute of Science, is developing a suite of tools for genomics annotation, in silico research and macromolecular structure analysis. In June, AlphaGene Inc. (Woburn, Mass.) announced a collaboration to use bioinformatics technology from Questar Bioinformatics Ltd. (Hyderabad) to mine AlphaGene's protein library. Questar will provide support for structure determination, pathway identification, and small molecule library development. The Indian government's reluctance to take risks and the scientific establishment's dependence on government have history.

First and foremost on Dhankar's list is putting a tight patent regime in place. "That includes adhering to the GATT agreement India has signed, and making sure there is enforcement. Often in India laws come into place but companies ignore them and things get tied up in courts for years," he said. Second, Dhankar said, is "fostering an environment of greater innovation and entrepreneurship. It is very difficult for someone to start up a new company today, but it can be done."

Finally, he said, India still has work to do to create a more business friendly environment, "including removal of certain duties and recourse to the courts." Although there is a modern legal system and independent judiciary, delays in moving cases through the courts limit their utility for resolving commercial disputes.

Other barriers include expensive and unreliable power, and high prices for transportation and real estate. While low wages more than compensate for high infrastructure costs, as the coun try becomes more integrated into the global economy, skilled anpower costs are likely to increase. The government is mobilizing to address the biggest obstacle to foreign investment in biomedical research: intellectual property protection. Under the TradeRelated Intellectual Property Rights (TRIPs) agreement, India must implement patent protection on pharmaceuticals and biotech products by 2005. In the runup to 2005, the nation's major biopharmaceutical companies are anticipating that the era of piggybacking on foreign i.p. is drawing to a close. Thus they are accelerating efforts to get bioequivalent versions of patented wellcharacterized recombinant proteins onto the market before the window closes.

"Come 2005, we will not be able to do any more recombinant proteins that are patented. You have four years to do whatever you like for the people of India," RDY's Reddy told a group of biotech and pharmaceutical executives Pushing clinical trial capacity establish 10 clinical centres and has prequalified physicians and hospitals capable of conducting trials for oncology, central nervous system, endocrinology, infectious diseases, internal medicine and cardiovascular drugs. It costs less to conduct trials in India than in developed countries, but the primary reason for working there is the availability of patients and investigators, according to Ferzaan Engineer, managing director of Quintiles India. "In absolute terms, certainly India is cost effective. But I'd add the time factor. If we can do a study in half the time, that is more important to a customer than us doing it 20 percent cheaper," he said.

With a population of 1 billion, the larger patient pools are an important factor. Engineer noted that the incidence of oral cancer and diabetes are far higher in India than in the U.S. and Europe. India is also a natural place to conduct trials of therapies for infectious diseases. AstraZeneca plc has established a laboratory and clinical research centre in Bangalore dedicated to discovery and clinical trials for diseases of relevance to the developing world, particularly tuberculosis. The company also operates a research foundation in India that conducts and sponsors basic molecular biology research. Similarly, Pfizer Inc.'s Indian subsidiary conducts clinical trials for its parent. - Steve Usdin.

The Drug Controller General of India recently announced that new regulations for reviewing applications to test investigational drugs will be released by the end of November. The regulatory regime, modelled on World Health Organization and International Conference on Harmonisation guidelines, is intended to increase the number and scope of international trials conducted in India. In addition to conducting bioequivalence trials of approved products and studies involving new indications for drugs approved in developed countries, companies have begun selecting India as the site to conduct human studies of new drugs. Indeed, according to Ranjit Roy Chaudhury, chairman of an Indian Council of Medical Research toxicology panel that advises the Drug Controller on applications to conduct clinical trials, India has been the site of the first human studies of at least six new molecular entities since 1996. Attracted by large patient populations, networks of academic medical centres, and genetically distinct population groups, international and domestic companies are establishing clinical research organizations in India. Since it started working in India in 1997, Quintiles Transnational Corp. (QTRN, Research Triangle Park, N.C.), has conducted more than 35 trials, primarily Phase III trials to support regulatory submissions in the U.S. and Europe. The company has helped products or when foreign companies are vying with domestic entities, according to Chandra Prakash, chairman of the Biotech Committee of the All India Biotech Association (AIBA).

"The commitment of finances in Indian biotechnology is risky because the biotechnology product approval process in India is even more complex and timeconsuming than in the U.S.," he said. "It is impossible to state if or when a given biotechnology product, which is approved in many countries, will get approval in India." Prakash is officer on special duty for the Punjab State Council of Science and Technology. The CII's National Task Force on Biotechnology has developed a series of recommendations for regulatory reform. Prime Minister Atal Bihari Vajpayee announced the government's intention to implement many of the steps suggested by CII and other business groups in September when he released a document outlining the nation's vision for biotechnology. The document commits the Department of Biotechnology to streamline approval procedures for biotech products and to finalize intellectual property laws and regu meeting in New Delhi last summer. He said that RDY plans to announce six other patented recombinant proteins over the next two years.

"India will adopt the spirit and the letter of intellectual property laws," Ranbaxy's Tempest. "Our plans are based on the assumption that we will be operating in a climate in which product patents will be enforced," he said. Improving the regulatory environment is another essential precondition for the emergence of a worldclass biotech sector, according to Jubilant's Chaudhri, who was one of the organizers of last summer's Confederation of Indian Industries (CII) conference on regulatory reform. She noted that the "pathway for approval of recombinant products is unclear, multiple government entities have overlapping jurisdiction, and there is ambiguity regarding the timelines for approvals for diverse types of biotechnology products." Approvals are sometimes complicated by conflicts of interest when government researchers are developing competing......?lations for product patents by 2003. Finally, over the last 15 years, the Department of Biotechnology has invested about $500 million in academic research and training. But ironically, the university and pharmaceutical company laboratories of the industrialized nations have benefited greatly from this investment, as the dearth of opportunities at home led generations of Indian scientists to go abroad.

The government is determined to stem the brain drain, in part by integrating its national laboratories and universities, which have been isolated from commerce by law and custom, according to Ragunath Mashelkar, director general of the Council of Scientific and Industrial Research. Under the new schemes, the government is putting its weight behind technology transfer into the private sector.

"The mindset of Indian academics has changed, now there is an orientation to transfer technology to industry," Manju Sharma, secretary of the Department of Biotechnology, told BioCentury.

Senior Writer Gita Kumar contributed to this report.

### *Institutes*

- Institution Affiliation Centre for Cellular & Molecular Biology (Hyderabad) CSIR
- Indian Institute of Chemical Technology (Hyderabad)

- CSIR Central Drug Research Institute (Lucknow)
- CSIR Centre for Biochemical Technology (New Delhi)
- CSIR Centre for Development of Advanced Computing (Pune)
- Ministry of Information Technology Indian Institute of Chemical Biology (Kolkata)
- CSIR Indian Institute of Science (Bangalore) Independent academic institution
- Indian Institute of Technology (Chennai, Delhi)
- Public academic institution Kanpur, Kharagpur, Mumbai)
- Industrial Toxicology Research Centre (Lucknow)
- CSIR International Centre for Genetic Engineering & United.

Nationsaffiliated Biotechnology (New Delhi) organization :

- National Brain Research Centre (New Delhi)
- Department of Biotechnology National Centre for Biological Sciences (Mumbai)
- Tata Institute of Fundamental Research National Chemical Laboratory (Pune)
- CSIR National Institute of Immunology (New Delhi).

### Hyderabad: A BT Hotbed

In addition to putting strong patent protections into effect, the key milestones for the Indian government will include implementing streamlined, transparent and predictable regulatory pathways. In the private sector, startups created to copy patented biologicals will be challenged to move into research, independently or under contract from U.S. and European companies, and for lowcost generic producers to make the transition to discoverybased endeavors. The Indian pharmaceutical industry, including domestic and export sales as well as contract services, totals about $5.5 billion, according to McKinsey and Co. "Assuming the industry, and to a lesser extent the government, can put in place solutions to three or four barriers, we see it growing to about $25 billion in 10 years," said McKinsey's Dhankar. But, he added, "we don't see that happening unless government and industry make some fairly significant India, a nation that has enthusiastically embraced and been transformed by information technology, is in the midst of a biotech frenzy. Newspapers and politicians hail "BT" as the savior for the dotbombed IT sector, and scores of computer software firms have announced the formation of bioinformatics spinoffs.

Hyderabad, capital of the southcentral state of Andhra Pradesh, competes for biotech preeminence with Bangalore, the country's cosmopolitan IT centre. Bangalore, capital of Karnataka in south India, boasts Biocon India Ltd., the first and largest biotech player in India, the Indian Institute of Science, scores of startups oriented to informatics and genomics, and a number of venture companies looking to diversify from IT to the life sciences. Hyderabad is home base for the Centre for Cellular and Molecular Biology (CCMB), the most entrepreneurial of the country's national laboratories, a large segment of the generic pharmaceutical industry, startup recombinant vaccine makers Shantha Biotechnics Pvt. Ltd. and Bharat Biotech International Ltd., and Dr. Reddy's Laboratory Ltd. (RDY), a multinational generics manufacturer that is planning to challenge the firstgeneration toptier biotechs with offpatent versions of blockbuster genetically engineered proteins.

Hyderabad is also ground zero for BThype. While its scores of biopharmaceutical companies and thousands of life sciences and chemistry graduate students highlight the reality and promise of Indian biotech, the city's Biotech Barber Shop, an insalu brious lowtech establishment in the city centre, exemplifies the hope that achieving bioprosperity is simply a matter of rebranding. On the outskirts of the clogged streets of Hyderabad and its twin city Secunderabad, on roads that wind through fields tilled by buffalopower and dusty villages, the state and national governments and the private sector are trying to recreate Silicon Valley's fusion of academic science and entrepreneurialism. The state's Genome Valley is being carved from the badlands 40 kilometres north of the city, where Andhra Pradesh is offering tax concessions and infrastructure support for companies to locate in Biotechnology Park.

The first occupant is Bharat Biotech, which was established by a group of Indian scientists living in the U.S. and led by Krishna Ella, a molecular biologist who moved to Hyderabad from Wisconsin in 1996. ICICI, one of the country's largest investment funds, has built a life sciences R&D incubator adjacent to the state's Biotechnology Park. ICICI has gone to great lengths to recreate the atmosphere of a California biotech campus at its Hyderabad Knowledge Park, equipping it with climatecontrolled modern plugandplay labs. But there also are contrasts with California that go well beyond the bright saris the grounds workers wear. ICICI's self sufficiency is perhaps unparalleled, including its own electricity substation, water and sewage plants. - Steve Usdin

## *India's Deal Flow*

Selected collaborations between Indian companies and institutions with companies in North America and Europe since the beginning of

1998. List excludes deals to distribute drugs and diagnostics inside India. Date Indian company/institution

Other company Deal Oct01 Wockhardt (BSE:532300) Rhein Biotech (NMarkt:RBO, Wockhardt is in the process of buying RBO's equity Maastricht, the Netherlands) holding in their 1996 Wockhardt Rhein Biopharm joint venture, which is producing a recombinant hepatitis B vaccine, BiovacB. Jun01 Questar AlphaGene (Woburn, Mass.) The companies will use Questar's bioinformatics technology to mine AlphaGene's protein library. Jun01 Ranbaxy Labs (BSE:500359) Vectura (Bath, U.K.) Ranbaxy's Ranbaxy B.V. subsidiary and Vectura will develop an oral controlledrelease drug delivery technology using Vectura's controlledrelease drug delivery solutions. May01 Zydus Cadila (BSE:532321)

Onconova Therapeutics Zydus Cadila entered into a joint venture with Onconova (Lawrenceville, N.J.) Therapeutics for collaborative oncogenomics research, manufacturing and marketing. May01 Dr. Reddy's Labs (NYSE:RDY; Novartis (SWX:NOVN; NVS) RDY granted NVS exclusive worldwide development and BSE:500124) marketing rights to its DRF 4158 insulin sensitizer to treat Type II diabetes. Jan01 Zydus Cadila (BSE:532321) Pantheco (Copenhagen, The parties will develop antibiotic compounds based on Denmark) existing classes of antibiotic drugs. Apr00 Tumkur Chemicals

Phytopharm (LSE:PYM, Tumkur will manufacture two PYM compounds derived from Godmanchester, U.K.) native Indian plants, P54 and P56. Apr00 Shantha Biotechnics Pfizer (PFE, New York, N.Y.) PFE obtained right of first refusal to become the exclusive comarketer for any new products developed by Shantha. Mar99 Proagro Hoechst Schering AgrEvo AgrEvo acquired Proagro, which specializes in seed (Berlin, Germany) breeding of rice, corn, cotton and oil seed rape, for an undisclosed amount. Feb99 Indian Institute of Science MitoKor (San Diego, Calif.) MitoKor will fund research at the Institute to synthesize intermediates for use in the company's programs. MitoKor retains exclusive rights to compounds developed in the program, and to certain synthetic processes. Sep98 Shanta Biotechnics American Diversified American Diversified will exclusively distribute Shanta's (Hickory, N.C.) ShanvacB hepatitis B vaccine in South America, Asia and Africa.

### Risk of Allergy from Genetically Engineered Products

Fears of risk of allergy, from products of Genetic Engineering (GE), have become a major issue in biosecurity considerations. Many proteins are immunogens and antigens, which elicit the production of different types of immunoglobulin (Ig) antibodies in mammalian systems. In response to the presence of immunogens and antigens, immunoglobulin M (IgM) antibodies form first but the quantity of the subsequently formed immunoglobulin G (IgG) antibodies is the highest of all. IgA are involved in the defence of the oral cavity. The function of IgD antibodies, which occur in small quantities, is not well understood. These antibodies fight infection in our body system.

For some unknown reasons, our immunological system reacts to certain proteins in an entirely different way, to produce another class of antibodies, the IgE. These are the allergens.

IgE molecules bind to mast cells in the internal and external linings of the body. The mast cells have large membrane bound organelles containing histamine. The binding of IgE to the mast cells causes the release of histamine triggering an inflammatory reaction, in the tissue layers containing mast cells. Externally, this manifests in the skin as rashes or weals, and internally the gastric lining becomes inflamed, often associated with stomach cramps. Bronchial allergies are caused mostly by inhalants. While most allergens are proteins, secondary metabolites (haptens), such as penicillin or parthenin, also can cause severe reactions. Haptens need to bind to a carrier protein to be allergenic. Usually this is an endogenous protein, already present in our body.

The best way to avoid allergy is to avoid contact with the allergen, basing on each individual's experience. Immunological treatment of allergy involves identifying the allergen through dermal tests and introducing it into the body system in small doses over a period of time. Slowly, the body system responds into producing more and more of IgG antibodies, which bind to and neutralise the allergen before large quantities of IgE are produced. Most of us are protected from an innumerable number of allergens by a similar natural process.

Immunological treatment of allergy is not always successful and is not feasible when an endogenous protein is involved, because the body does not produce antibodies to its own protein (except in the case of autoimmune diseases). Proteins in conventional food items such as fish, eggs, milk, peanuts, Brazil nuts, certain varieties of rice, cucumbers, mushrooms and many others cause allergy in different people and so do certain drugs. Most products of GE are not potentially more allergenic than their conventional counterparts. The risk of allergy needs to be considered only when a GE food or drug contains one or more new proteins, not present in the isogenic variety, coded by the introduced genes. For example, Bt potato tuber contains the Bt protein, which was found to be safe. When there are no new proteins in a GE product, the question of allergy has no relevance. If some one is allergic to conventional peanuts, the same will happen with GE peanuts as well. Extensive tests are being conducted to check for allergenicity of GE products. The following considerations need to be kept in mind in this regard: a) tests use only models of known allergenic proteins and do not involve all proteins or probable non-protein allergens, b) it is near impossible to test for all the proteins and non-proteins in a product for the potential of allergy, c) there is no single allergen that can cause allergy in all the people even in a single household, and d) people tend to be allergic to certain substances only during a period of their life and not all through.

Allergy is neither a new nor a universal issue. It is a problem for some individuals and it is not equivalent to an infection. Any chemical substance in the environment or that is ingested into our body system, can be allergenic. Most allergies disappear as mysteriously as they developed. We have not stopped the production of fish, eggs, milk, peanuts, etc., for the reason they cause allergy in some people. What is needed is a rational attitude with concern for larger benefits.

The happenings in Andhra Pradesh to malign agbiotech, which are sure to follow in other states, are disturbing. More alarming are the signs of the campaign getting physical to destroying standing crops and the personal risk to the farmers themselves. This is where the State Government should get ready to step in and deal with the situation as a Law and Order issue.

The arguments for and against biotechnology must be based in science. In India public awareness of issues related to biotechnology is abysmal. Even educated public, including some biologists, do not have a clear perception of the benefits and risks of agbiotech. The general policy of the Government of India is pro-technology, but its departments have not been involved in the area of public awareness and education programmes in any appreciable manner. The Governments of the States are busy releasing policy documents that focus on biotech business, with products nowhere in sight. In India the state of biotechnology education itself is in a deplorable mess with hype and reality at opposite poles. The biotech industry has done precious little to educate even farmers using their products, let alone the general public. The media only care for sensational news like demonstrations and vandalization. They have not been enthusiastic even about getting educated themselves, and are not equal to the task of playing any constructive role in enhancing public understanding. While some Muslim religious bodies and the Vatican have come in support of biotechnology, religious authorities in general are in the dark. The activists have been taking advantage of this huge information gap, and mix up ethical, economic and political issues, to spread misinformation and to create scare in the public mind, from the emotive and sentimental platform.

It is high time that a far-reaching programme of public awareness and education is set in motion in India. Countering the sustained attack on agbiotech requires a considerable financial support. While the activists are functioning with a lot of financial backing from organizations with vested interest in maligning biotechnology, those who are pleading for a chance for agbiotech to prove or disprove itself in course of time, have no resources to speak of.

The agbiotech industry has not fully realized its responsibility to help the cause of disseminating factual information and to assuage the doubts and fears of the public. Conducting workshops and seminars, sparsely supported by one or two agbiotech companies or the Department of Biotechnology, has its own undeniable benefits but it does not go much beyond the urban scene. The agbiotech industry should realize the seriousness of the consequences of anti-biotech campaign and its own moral responsibility in the matter, and come forward to fully support public awareness and education programmes. I made a case for such a realization of the industry's responsibility, at a Syngenta Lecture in Basel, last February.

If an organization or a group of individuals supportive of biotechnology conducts a programme with funding from the industry, they immediately win the label 'toadies of the industry', whereas the anti-biotech lobby, that receives huge funding from mostly European Luddite greens, parade as the protectors of the environment and patriots protecting their countries form the marauding multi-nationals. Both the supporters of technology and the industry should join hands and work together in the common cause of biotechnology. Simply because an organization receives financial support from the industry, should not mean that the organization sold itself and blindly supports every product disregarding ethics of science and public interest, which are as important as the technology itself.

A better option would be for the various agbiotech companies to form a neutral body that receives and pools up funds from the Indian industry and supports public education activities of proactive organizations and educational institutions. At the international level, there are such bodies like Crop Life International, receiving funds from the industry and in turn support biotech related activities. A national level body would minimize the procedural and time hassles.

Can AgBioWorld Foundation work toward this end?

All this will take time, but something must be done on an urgent basis to counter the anti-agbiotech campaign first in Andhra Pradesh and subsequently in other parts of India. In the Indian epic Ramayana, Kumbhakarna, who would go into a comatose sleep months on, woke up when his services were needed. Would the agbiotech industry wake up and support the cause at least now?

### *Preparation to Face Bioterrorism*

India is among the most vulnerable countries that run the risk of facing threats of bioterrorism. A lot of R&D and groundwork has to be planned and carried out speedily, for the vulnerable countries to offer even a semblance of fight against bioterrorism. In the current absence of awareness and preparedness, we need to examine what the developed world has been doing to face bioterrorism.

A number of recommendations, based on a very comprehensive survey, to deal with chemical and biological terrorism and to increase civilian medical response, appeared in the publication "Improving Civilian Medical Response to Chemical or Biological Terrorist Incidents: Interim Report on Current Capabilities (1998)".

This report was prepared by the Committee on R & D Needs for Improving Civilian Medical Response to Chemical and Biological Terrorism Incidents, Division of Health Science Policy, Institute of Medicine, Board on Environmental Studies and Toxicology, National Research Council, in the US. While we cannot directly import the western models, which are aimed at improving an existing and reasonably effective preparedness, they will certainly help in planning our own strategies, basing on our ground realities.

A summary of the 10 R&D needs and eight recommendations from this report is given here. Developing countries do not have the organisational infrastructure basis comparable to that in North America or Europe (mentioned in here), although there are some governmental and non-governmental organisations, involved in R&D in public health, drug development and medical issues. These efforts need to be upgraded to suit each country's requirements to face bioterrorism and/or natural calamities. The R&D needs and recommendations presented here may seem to be simple at the first glance, but they are loaded with prescription for extensive and expensive work, on several fronts simultaneously and each one of the 18 statements requires an elaborate explanation, to put it into action. They would do a lot of good to gear up the potential of a country to manage disease in its day-to-day occurrence and prepare as never before to face natural calamities such as cyclones, floods and earthquakes, which epidemics follow in the wake.

### Research and Development Needs

1. A system is needed to ensure that medical facilities receive information on actual, suspected, and potential terrorist activity. Research may be necessary to determine what should be communicated, to whom it should be communicated and even whether the system should vary by state and city, but it must include links to the law enforcement community.
2. The civilian medical community must find ways to adapt the many new and emerging detection technologies to the spectrum of chemical and biological warfare agents. First responders, emergency medical personnel, and public safety officials, all need improved instrumentation for detecting and identifying chemical and biological agents in both the environment and in clinical samples from patients. The watchwords are simplicity, speed, cost, sensitivity and specificity. The key to widespread purchase and uses lies with identifying a wide spectrum of toxic substances, including but not limited to military agents.
3. Work on symptom-based tools for identifying unknown toxic agents, including but not limited to, military chemical weapons, is an area where benefits may extend well beyond response to terrorist acts.
4. Complete information is needed on the toxicity and adverse health effects that could result from acute exposure to low levels of agents, especially in sensitive populations, such as the young, the elderly and those in ill health. This information is necessary to develop guidelines (for example, susceptible human exposure levels) for safe and effective evacuation, decontamination, and other protective action.
5. Methods are needed for rapid, effective, and inexpensive decontamination of large groups of personnel, equipment and environment.
6. Approaches to treatment are needed that have utility beyond terrorism or chemical and biological warfare. Vaccines, or drugs aimed at families of pathogens or toxins, substances to bind toxic molecules before they reach their site of action and perhaps even existing drugs and other chemicals that can serve as expedient treatment (for example, anticholinergics other than atropine), are to be identified.
7. Complete information is needed on possible interactions of antidotes and therapeutic drugs with anti-hypertensives, psychotherapeutics, anti-inflammatory compounds, immunosuppressants, and other medications in widespread public use.

8. There is a need for evaluation of interventions for preventing or ameliorating adverse psychological effects in emergency workers, victims, and near-victims. Examination of the Japanese experience following the release of sarin on the Tokyo subway, other acts of terrorism (recent threat of anthrax in the US), and unintentional releases of toxic chemicals (the Bhopal gas tragedy) would be especially valuable.
9. Information is needed on risk assessment/threat perception by individuals and groups, and on risk communication by public officials, especially the roles of both the mass media and the Internet in the transmission of anxiety (or confidence). Some information is available of pollutants and toxic waste, but there is little or no systematically collected data on fears and anxieties related to the possibility of purpose fully introduced disease.
10. Standardised protocols for follow-up of first responders, healthcare providers and victims are required, for improving care of those individuals, for improving medical response to future incidents, and for improving risk assessments.

### *Interim Recommendations*

The Committee considered it irresponsible to focus solely on technology R&D that requires elaborate and meticulous planning and is both time consuming and expensive. Hence, the Committee made eight recommendations involving potentially simpler, faster or less expensive mechanisms than R&D of new technology. These are slightly modified here for the use of the Developing Countries, which should make a beginning with these recommendations and also take up the R&D measures, simultaneously.

1. Provide financial support for improvements in state and local surveillance infrastructure, such as poison control centres and communicable disease programmes.
2. Survey major metropolitan hospitals for supplies of antidotes, drugs, ventilators, personal protective equipment, decontamination capacity, mass-casualty planning and training, isolation rooms for infectious disease, and familiarity of staff with the effects and treatment of chemical and biological weapons.
3. Encourage the governmental and private agencies engaged in health and medical R&D to share their information on diseases and drugs and on the location and owners of dangerous biological materials. State health departments in turn should be encouraged, by education or training, on the effects of agents and medical responses required, to add infections by these materials to their lists of reportable diseases.

4. Provide support to the Army's efforts to test commercial personal protective equipment for protection against nerve and vesicants.
5. Convene discussions among the appropriate agencies on the use of investigational products in mass-casualty situation and on acceptable proof of efficacy for products where clinical trials are not ethical or are otherwise impossible.
6. Develop incentives for hospitals, both public and private, to be ambulance receiving hospitals, to stockpile nerve-agent antidotes and selected antitoxins and put them in the hands of first responders, by changing laws if needed, to purchase appropriate personal protective equipment and expandable decontamination facilities and train emergency department personnel in their use.
7. Provide for state and central training initiatives with a programme to incorporate existing information on possible chemical and biological terror agents and their treatment into the manuals and reference libraries of first responders, emergency departments and poison control centres. Professional societies and journal publishers should be recruited to help in this effort.
8. Intensify Public Health Service efforts to organise and equip Urban Medical Strike Teams, in high-risk cities throughout the country. Although these teams are to be primarily designed to cope up with terrorism, using local personnel and resources, they also increase the community's general ability to cope with industrial accidents and other mass-casualty events.

# Chapter 6

# Evolutionary Approaches to Computational Biology

Computational biology is an interdisciplinary field that applies the techniques of computer science, applied mathematics, statistics and engineering to address biological problems. Some of these problems involve the development of computational and statistical data analysis methods and in developing mathematical modelling and computational simulation techniques. By these means it addresses scientific research topics with their theoretical and experimental questions without a laboratory. It is connected to the following fields:

- Computational biomodeling, a field concerned with building computer models of biological systems.
- Bioinformatics, which applies algorithms and statistical techniques to the interpretation, classification and understanding of biological datasets. These typically consist of large numbers of DNA, RNA, or protein sequences. Sequence alignment is used to assemble the datasets for analysis. Comparisons of homologous sequences, gene finding, and prediction of gene expression are the most common techniques used on assembled datasets; however, analysis of such datasets have many applications throughout all fields of biology.
- Mathematical biology aims at the mathematical representation, treatment and modelling of biological processes, using a variety of applied mathematical techniques and tools.
- Computational genomics, a field within genomics which studies the genomes of cells and organisms. High-throughput genome sequencing produces lots of data, which requires extensive post-

processing (genome assembly) and uses DNA microarray technologies to perform statistical analyses on the genes expressed in individual cell types. This can help find genes of interest for certain diseases or conditions. This field also studies the mathematical foundations of sequencing. Advances in many areas of genomics research are heavily rooted in engineering technology, from the capillary electrophoresis units used in large-scale DNA sequencing projects.

- Molecular modelling, which consists of modelling the behaviour of molecules of biological importance.
- Protein structure prediction and structural genomics, which attempt to systematically produce accurate structural models for three-dimensional protein structures that have not been determined experimentally.
- Computational biochemistry and biophysics, which make extensive use of structural modelling and simulation methods such as molecular dynamics and Monte Carlo method-inspired Boltzmann sampling methods in an attempt to elucidate the kinetics and thermodynamics of protein functions.

## Modelling of Biological Systems

Modelling biological systems is a significant task of systems biology and mathematical biology. Computational systems biology aims to develop and use efficient algorithms, data structures, visualization and communication tools to orchestrate the integration of large quantities of biological data with the goal of computer modelling. It involves the use of computer simulations of biological systems, like cellular subsystems (such as the networks of metabolites and enzymes which comprise metabolism, signal transduction pathways and gene regulatory networks) to both analyse and visualize the complex connections of these cellular processes.

It is also directly associated with bioinformatics and computational biology. Artificial life or virtual evolution attempts to understand evolutionary processes via the computer simulation of simple (artificial) life forms.

It is understood that an unexpected emergent property of a complex system is a result of the interplay of the cause-and-effect among simpler, integrated parts. Biological systems manifest many important examples of emergent properties in the complex interplay of components. Traditional study of biological systems requires reductive methods in which quantities of data are gathered by category, such as concentration over time in response to a certain stimulus. Computers are critical to analysis and modelling of these data. The goal is to create accurate real-time models of a system's response to environmental and internal stimuli, such as a model of a cancer cell in order to find weaknesses in its signalling pathways, or modelling of ion channel mutations to see effects on cardiomyocytes and in turn, the function of a beating heart.

A monograph on this topic summarizes an extensive amount of published research in this area up to 1987, including subsections in the following areas: computer modelling in biology and medicine, arterial system models, neuron models, biochemical and oscillation networks, quantum automata, quantum computers in molecular biology and genetics, cancer modelling, neural nets, genetic networks, abstract relational biology, metabolic-replication systems, category theory applications in biology and medicine, automata theory, cellular automata, tessallation models and complete self-reproduction, chaotic systems in organisms, relational biology and organismic theories.

This published report also includes 390 references to peer-reviewed articles by a large number of authors.

### Standards

By far the most widely accepted standard format for storing and exchanging models in the field is the Systems Biology Markup Language (SBML) The SBML.org website includes a guide to many important software packages used in computational systems biology. Other markup languages with different emphases include BioPAX and CellML.

## Cellular Model

Creating a cellular model has been a particularly challenging task of systems biology and mathematical biology. It involves the use of computer simulations of the many cellular subsystems such as the networks of metabolites and enzymes which comprise metabolism, signal transduction pathways and gene regulatory networks to both analyse and visualize the complex connections of these cellular processes.

The complex network of biochemical reaction/transport processes and their spatial organization make the development of a predictive model of a living cell a grand challenge for the 21st century.

E-Cell Project aims "to make precise whole cell simulation at the molecular level possible".

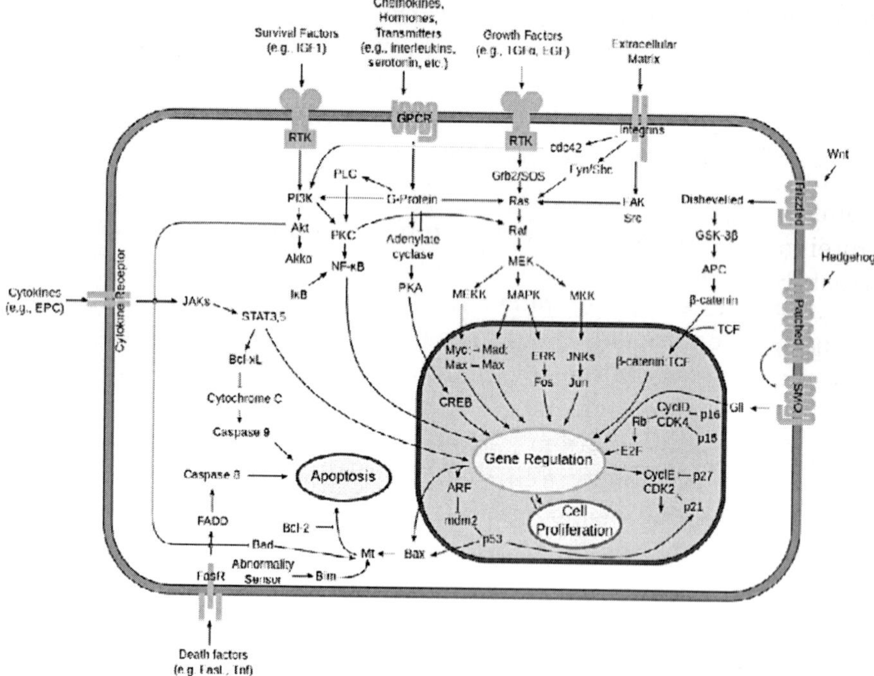

Membrane computing is the task of modelling specifically a cell membrane.

## Protein Folding

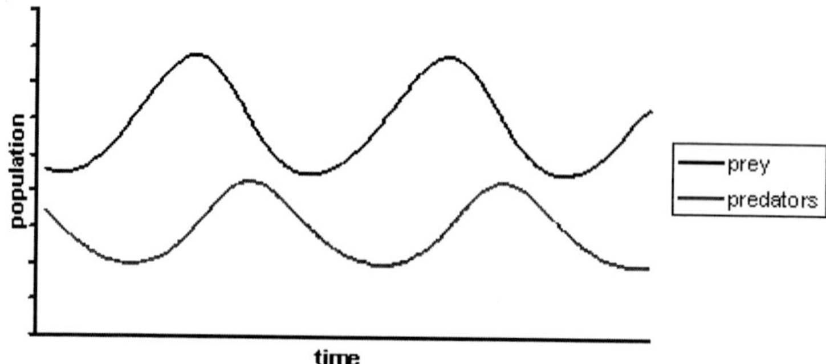


Protein structure prediction is the prediction of the three-dimensional structure of a protein from its amino acid sequence — that is, the prediction of its secondary, tertiary, and quaternary structure from its primary structure. Structure prediction is fundamentally different from the inverse problem of protein design. Protein structure prediction is one of the most important goals pursued by bioinformatics and theoretical chemistry; it is highly important in medicine (for example, in drug design) and biotechnology (for example, in the design of novel enzymes). Every two years, the performance of current methods is assessed in the CASP experiment (Critical Assessment of Techniques for Protein Structure Prediction).

### Secondary Structure

Secondary structure prediction is a set of techniques in bioinformatics that aim to predict the local secondary structures of proteins and RNA sequences based only on knowledge of their primary structure-amino acid or nucleotide sequence, respectively. For proteins, a prediction consists of assigning regions of the amino acid sequence as likely alpha helices, beta strands (often noted as "extended" conformations), or turns. The success of a prediction is determined by comparing it to the results of the DSSP algorithm applied to the crystal structure of the protein; for nucleic acids, it may be determined from the hydrogen bonding pattern. Specialized algorithms have been developed for the detection of specific well-defined patterns such as transmembrane helices and coiled coils in proteins, or canonical microRNA structures in RNA.

The best modern methods of secondary structure prediction in proteins reach about 80% accuracy; this high accuracy allows the use of the predictions in fold recognition and ab initio protein structure prediction, classification of structural motifs, and refinement of sequence alignments. The accuracy of current protein secondary structure prediction methods is assessed in weekly benchmarks such as LiveBench and EVA.

### Background

Early methods of secondary structure prediction, introduced in the 1960s and early 1970s, focused on identifying likely alpha helices and were based mainly on helix-coil transition models. Significantly more accurate predictions that included beta sheets were introduced in the 1970s and relied on statistical assessments based on probability parameters derived from known solved structures. These methods, applied to a single sequence, are typically at most about 60-65% accurate, and often underpredict beta sheets. The evolutionary conservation of secondary structures can be exploited by simultaneously assessing many homologous sequences in a multiple sequence alignment, by calculating the net secondary structure propensity of an aligned column of amino acids. In concert with larger databases of known protein structures and modern machine learning methods such as neural nets and support vector machines, these methods can achieve up 80% overall accuracy in globular proteins. The theoretical upper limit of accuracy is around 90%, partly due to idiosyncrasies in DSSP assignment near the ends of secondary structures, where local conformations vary under native conditions but may be forced to assume a single conformation in crystals due to packing constraints. Limitations are also imposed by secondary structure prediction's inability to account for tertiary structure; for example, a sequence predicted as a likely helix may still be able to adopt a beta-strand conformation if it is located within a beta-sheet region of the protein and its side chains pack well with their neighbours. Dramatic conformational changes related to the protein's function or environment can also alter local secondary structure.

### Chou-Fasman Method

The Chou-Fasman method was among the first secondary structure prediction algorithms developed and relies predominantly on probability parameters determined from relative frequencies of each amino acid's appearance in each type of secondary structure. The original Chou-Fasman parameters, determined from the small sample of structures solved in the mid-1970s, produce poor results compared to modern methods, though the parameterization has been updated since it was first published. The Chou-Fasman method is roughly 50-60% accurate in predicting secondary structures.

### GOR Method

The GOR method, named for the three scientists who developed it- *G*arnier, *O*sguthorpe, and *R*obson- is an information theory-based method developed not long after Chou-Fasman. It uses a more powerful probabilistic techniques of Bayesian inference. The method is a specific optimized application of mathematics and algorithms developed in a series of papers by Robson and colleagues, eg. and). The GOR method is capable of continued extension by such principles, and has gone through several versions. The GOR method takes into account not only the probability of each amino acid having a particular secondary structure, but also the conditional probability of the amino acid assuming each structure given the contributions of its neighbours (it does not assume that the neighbours have that same structure). The approach is both more sensitive and more accurate than that of Chou and Fasman because amino acid structural propensities are only strong for a small number of amino acids such as proline and glycine. Weak contributions from each of many neighbours can add up to strong effect overall. The original GOR method was roughly 65% accurate and is dramatically more successful in predicting alpha helices than beta sheets, which it frequently mispredicted as loops or disorganized regions.. Later GOR methods considered also pairs of amino acids, significantly improving performance. The major difference from the following technique is perhaps that the weights in an implied network of contributing terms are assigned *a priori*, from statistical analysis of proteins of known structure, not by feedback to optimize agreement with a training set of such.

### Machine Learning

Neural network methods use training sets of solved structures to identify common sequence motifs associated with particular arrangements of secondary structures. These methods are over 70% accurate in their predictions, although beta strands are still often underpredicted due to the lack of three-dimensional structural information that would allow assessment of hydrogen bonding patterns that can promote formation of the extended conformation required for the presence of a complete beta sheet.

Support vector machines have proven particularly useful for predicting the locations of turns, which are difficult to identify with statistical methods. The requirement of relatively small training sets has also been cited as an advantage to avoid overfitting to existing structural data.

Extensions of machine learning techniques attempt to predict more fine-grained local properties of proteins, such as backbone dihedral angles in unassigned regions. Both SVMs and neural networks have been applied to this problem.

### Other Improvements

It is reported that in addition to the protein sequence, secondary structure formation depends on other factors. For example, it is reported that secondary structure tendencies depend also on local environment, solvent accessibility of residues, protein structural class, and even the organism from which the proteins are obtained. Based on such observations, some studies have shown that secondary structure prediction can be improved by addition of information about protein structural class, solvent accessibility and also contact number of residues.

Sequence covariation methods rely on the existence of a data set composed of multiple homologous RNA sequences with related but dissimilar sequences. These methods analyse the covariation of individual base sites in evolution; maintenance at two widely separated sites of a pair of base-pairing nucleotides indicates the presence of a structurally required hydrogen bond between those positions. The general problem of pseudoknot prediction has been shown to be NP-complete.

## Tertiary Structure

The practical role of protein structure prediction is now more important than ever. Massive amounts of protein sequence data are produced by modern large-scale DNA sequencing efforts such as the Human Genome Project. Despite community-wide efforts in structural genomics, the output of experimentally determined protein structures—typically by time-consuming and relatively expensive X-ray crystallography or NMR spectroscopy—is lagging far behind the output of protein sequences. The protein structure prediction remains an extremely difficult and unresolved undertaking. The two main problems are calculation of protein free energy and finding the global minimum of this energy. A protein structure prediction method must explore the space of possible protein structures which is astronomically large. These problems can be partially bypassed in "comparative" or homology modelling and fold recognition methods, when the search space is pruned by the assumption that the protein in question adopts a structure that is close to the experimentally determined structure of another homologous protein. On the other hand, the *de novo* or ab initio protein structure prediction methods must explicitly resolve these problems.

### Ab Initio Protein Modelling

*Ab initio*- or *de novo*- protein modelling methods seek to build three-dimensional protein models "from scratch", i.e., based on physical principles rather than (directly) on previously solved structures. There are many possible procedures that either attempt to mimic protein folding or apply some stochastic method to search possible solutions (i.e., global optimization of a suitable energy function). These procedures tend to require vast computational resources, and have thus only been carried out for tiny proteins. To predict protein structure *de novo* for larger proteins will require better algorithms and larger computational resources like those afforded by either powerful supercomputers (such as Blue Gene or MDGRAPE-3) or distributed computing (such as Folding@home, the Human Proteome Folding Project and Rosetta@Home). Although these computational barriers are vast, the potential benefits of structural genomics (by predicted or experimental methods) make *ab initio* structure prediction an active research field.

As an intermediate step towards predicted protein structures, contact map predictions have been proposed.

### Comparative Protein Modelling

Comparative protein modelling uses previously solved structures as starting points, or templates. This is effective because it appears that although the number of actual proteins is vast, there is a limited set of tertiary structural motifs to which most proteins belong. It has been suggested that there are only around 2,000 distinct protein folds in nature, though there are many millions of different proteins.

These methods may also be split into two groups :

***Homology Modelling :*** is based on the reasonable assumption that two homologous proteins will share very similar structures. Because a protein's fold is more evolutionarily conserved than its amino acid sequence, a target sequence can be modelled with reasonable accuracy on a very distantly related template, provided that the relationship between target and template can be discerned through sequence alignment. It has been suggested that the primary bottleneck in comparative modelling arises from difficulties in alignment rather than from errors in structure prediction given a known-good alignment. Unsurprisingly, homology modelling is most accurate when the target and template have similar sequences.

***Protein Threading:*** scans the amino acid sequence of an unknown structure against a database of solved structures. In each case, a scoring function is used to assess the compatibility of the sequence to the structure, thus yielding possible three-dimensional models. This type of method is also known as 3D-1D fold recognition due to its compatibility analysis between three-dimensional structures and linear protein sequences. This method has also given rise to methods performing an inverse folding search by evaluating the compatibility of a given structure with a large database of sequences, thus predicting which sequences have the potential to produce a given fold.

## Side Chain Geometry Prediction

Accurate packing of the amino acid side chains represents a separate problem. Methods that specifically address the problem of predicting side chain geometry include dead-end elimination and the self-consistent mean field methods. The side chain conformations with low energy are usually determined on the rigid polypeptide backbone and using a set of discrete side chain conformations known as "rotamers" or a "conformational isomerism". The methods attempt to identify the set of rotamers that minimize the model's overall energy.

These methods use rotamer libraries, the collections of rotamers (favourable multi-angle conformations) for each residue type in proteins. Rotamer libraries may contain information about the conformation, its frequency, and the variance about mean dihedral angles, which can be used in sampling. Rotamer libraries are derived from structural bioinformatics or other statistical analysis of side-chain conformations in known experimental structures of proteins, such as by clustering the observed conformations for tetrahedral carbons near the staggered (60°, 180°, -60°) values. Rotamer libraries can be backbone-independent, secondary-structure-dependent, or backbone-dependent. Backbone-independent rotamer libraries make no reference to backbone conformation, and are calculated from all available side chains of a certain type (for instance, the first example of a rotamer library, done by Ponder and Richards at Yale in 1987). Secondary-structure-dependent libraries present different dihedral angles and/or rotamer frequencies for $\alpha$-helix, $\beta$-sheet, or coil secondary structures. Backbone-dependent rotamer libraries present conformations and/or frequencies dependent on the local backbone conformation as defined by the backbone dihedral angles $\phi$ and $\psi$, regardless of secondary structure. The modern versions of these "libraries" as used in most software are presented as multidimensional distributions of probability or frequency, where the peaks correspond to the dihedral-angle conformations considered as individual rotamers in the lists. Some versions are especially sensitive to the prohibited regions in that conformational space and are used primarily for structure validation, while others emphasize relative frequencies in the favourable regions and are the form used primarily for structure prediction, such as the Dunbrack rotamer "libraries".

The side chain packing methods are most useful for analysing the protein's hydrophobic core, where side chains are more closely packed; they have more difficulty addressing the looser constraints and higher flexibility of surface residues, which often occupy multiple rotamer conformations rather than just one.

### Prediction of Structural Classes

Statistical methods have been developed for predicting structural classes of proteins based on their amino acid composition, pseudo amino acid composition and functional domain composition.

### Quaternary Structure

In the case of complexes of two or more proteins, where the structures of the proteins are known or can be predicted with high accuracy, protein–protein docking methods can be used to predict the structure of the complex. Information of the effect of mutations at specific sites on the affinity of the complex helps to understand the complex structure and to guide docking methods.

***Software:*** Modeller is a popular software tool for producing homology models using methodology derived from NMR spectroscopy data processing. SwissModel provides an automated web server for basic homology modelling. I-Tasser is the best server for protein structure prediction according to the 2006-2008 CASP experiments (CASP7 and CASP8).

HHpred, bioinfo.pl, Robetta, and Phyre are widely used servers for protein structure prediction. HHsearch is a free software package for protein threading and remote homology detection. Raptor (software) is a protein threading software that is based on integer programming. The basic algorithm for threading is described in and is fairly straightforward to implement. Abalone is a Molecular Dynamics program for folding simulations with explicit or implicit water models.

TIP is a knowledgebase of Structfast models and precomputed similarity relationships between sequences, structures, and binding sites. Several distributed computing projects concerning protein structure prediction have also been implemented, such as the Folding@home, Rosetta@home, Human Proteome Folding Project, Predictor@home, and TANPAKU.

The Foldit program seeks to investigate the pattern-recognition and puzzle-solving abilities inherent to the human mind in order to create more successful computer protein structure prediction software.

Computational approaches provide a fast alternative route to antibody structure prediction. Recently developed antibody $F_V$ region high resolution structure prediction algorithms, like RosettaAntibody, have been shown to generate high resolution homology models which have been used for successful docking. Reviews of software for structure prediction can be found at. The progress and challenges in protein structure prediction has been reviewed in Zhang 2008.

### Automatic Structure Prediction Servers

CASP, which stands for Critical Assessment of Techniques for Protein Structure Prediction, is a community-wide experiment for protein structure prediction taking place every two years since 1994. CASP provides users and research groups with an opportunity to assess the quality of available methods and automatic servers for protein structure prediction. Official results for automatic structure prediction servers in the CASP7 benchmark (2006) are discussed by Battey *et al.*. Official CASP8 results are available for automatic servers and for human and server predictors. Unofficial results for automatic servers of the 2008 CASP8 benchmark are summarized on several lab websites and ranked according to slightly varying criteria: Zhang lab, Grishin lab, McGuffin lab, Baker lab, and Cheng lab.

## Human Biological Systems

### *Brain Model*

The Blue Brain Project is an attempt to create a synthetic brain by reverse-engineering the mammalian brain down to the molecular level. The aim of the project, founded in May 2005 by the Brain and Mind Institute of the *École Polytechnique* in Lausanne, Switzerland, is to study the brain's architectural and functional principles. The project is headed by the Institute's director, Henry Markram. Using a Blue Gene supercomputer running Michael Hines's NEURON software, the simulation does not consist simply of an artificial neural network, but involves a partially biologically realistic model of neurons. It is hoped by its proponents that it will eventually shed light on the nature of consciousness. There are a number of sub-projects, including the Cajal Blue Brain, coordinated by the Supercomputing and Visualization Centre of Madrid (CeSViMa), and others run by universities and independent laboratories in the UK, U.S., and Israel.

### *Model of the Immune System*

The last decade has seen the emergence of a growing number of simulations of the immune system.

### *Tree Model*

The simulated growth of plants is a significant task in of systems biology and mathematical biology, which seeks to reproduce plant morphology with computer software. Electronic trees (e-trees) usually use L-systems to simulate growth. L-systems are very important in the field of complexity science and A-life. A universally accepted system for describing changes in plant morphology at the cellular or modular level has yet to be devised. The most widely implemented tree-generating algorithms are described in the papers "Creation and Rendering of Realistic Trees", and Real-Time Tree Rendering.

The realistic modelling of plant growth is of high value to biology, but also for computer games.

## Theory + Algorithms

A biologist, Aristid Lindenmayer (1925–1989) worked with yeast and filamentous fungi and studied the growth patterns of various types of algae, such as the blue/green bacteria *Anabaena catenula*. Originally the L-systems were devised to provide a formal description of the development of such simple multicellular organisms, and to illustrate the neighbourhood relationships between plant cells.

Later on, this system was extended to describe higher plants and complex branching structures. Central to L-systems, is the notion of rewriting, where the basic idea is to define complex objects by successively replacing parts of a simple object using a set of rewriting rules or productions. The rewriting can be carried out recursively. L-Systems are also closely related to Koch curves.

### Environmental Interaction

A challenge for plant simulations is to consistently integrate environmental factors, such as surrounding plants, obstructions, water and mineral availability, and lighting conditions is to build virtual/ environments with as many parameters as computationally feasible, thereby, not only simulating the growth of the plant, but also the environment it is growing within, and, in fact, whole ecosystems. Changes in resource availability influence plant growth, which in turn results in a change of resource availability. Powerful models and powerful hardware will be necessary to effectively simulate these recursive interactions of recursive structures.

### Software

- Branching: L-system Tree0A Java applet and its source code (open source) of the botanical tree growth simulation using the L-system.
- Arbaro-opensource
- Treal-opensource
- L-arbor
- AMAP Genesis 3.0 -from CIRDAD
- ONETREE-Accompanying the CDROM is a $CO_2$ meter that plugs into your local serial port. It is this that controls the growth rate of the trees. It is the actual carbon dioxide level right at your computer that controls the growth rate of these virtual trees.
- Powerplant.

Electronic trees (e-trees) usually use L-systems to simulate growth. L-systems are very important in the field of complexity science and A-life. A universally accepted system for describing changes in plant morphology at the cellular or modular level has yet to be devised.

The most widely implemented tree generating algorithms are described in the papers "Creation and Rendering of Realistic Trees", and Real-Time Tree Rendering.

## Ecological Models

Ecosystem models, or ecological models, are mathematical representations of ecosystems. Typically they simplify complex foodwebs down to their major components or trophic levels, and quantify these as either numbers of organisms, biomass or the inventory/ concentration of some pertinent chemical element (for instance, carbon or a nutrient species such as nitrogen or phosphorus).

### *Complexity*

Ecosystem models are a development of theoretical ecology that aim to characterise the major dynamics of ecosystems, both to synthesise the understanding of such systems and to allow predictions of their behaviour (in general terms, or in response to particular changes).

Because of the complexity of ecosystems (in terms of numbers of species/ecological interactions), ecosystem models typically simplify the systems they are studying to a limited number of pragmatic components. These may be particular species of interest, or may be broad functional types such as autotrophs, heterotrophs or saprotrophs. In biogeochemistry, ecosystem models usually include representations of non-living "resources" such as nutrients, which are consumed by (and may be depleted by) living components of the model.

This simplification is driven by a number of factors:

- Ignorance: while understood in broad outline, the details of a particular foodweb may not be known; this applies both to identifying relevant species, and to the functional responses linking them (which are often extremely difficult to quantify)
- Computation: practical constraints on simulating large numbers of ecological elements; this is particularly true when ecosystem models are embedded within other spatially-resolved models (such as physical models of terrain or ocean bodies, or idealised models such as cellular automata or coupled map lattices)
- Understanding: depending upon the nature of the study, complexity can confound the analysis of an ecosystem model; the more interacting components a model has, the less straightforward it is to extract and separate causes and consequences; this is compounded when uncertainty about components obscures the accuracy of a simulation.

### *Structure*

The process of simplification described above typically reduces an ecosystem to a small number of state variables. Depending upon the system under study, these may represent ecological components in terms of numbers of discrete individuals or quantify the component more continuously as a measure of the total biomass of all organisms of that type, often using a common model currency (e.g. mass of carbon per unit area/volume).

The components are then linked together by mathematical functions that describe the nature of the relationships between them. For instance, in models which include predator-prey relationships, the two components are usually linked by some function that relates total prey captured to the populations of both predators and prey. Deriving these relationships is often extremely difficult given habitat heterogeneity, the details of component behavioural ecology (including issues such as perception, foraging behaviour), and the difficulties involved in unobtrusively studying these relationships under field conditions.

Typically relationships are derived statistically or heuristically. For example, some standard functional forms describing these relationships are linear, quadratic, hyperbolic or sigmoid functions. The latter two are known in ecology as type II and type III responses, named by C. S. Holling in early, groundbreaking work on predation in mammals. Both describe relationships in which a linkage between components saturates at some maximum rate (e.g. above a certain concentration of prey organisms, predators cannot catch any more per unit time). Some ecological interactions are derived explicitly from the biochemical processes that underlie them; for instance, nutrient processing by an organism may saturate because of either a limited number of binding sites on the organism's exterior surface or the rate of diffusion of nutrient across the boundary layer surrounding the organism. After establishing the components to be modelled and the relationships between them, another important factor in ecosystem model structure is the representation of space used. Historically, models have often ignored the confounding issue of space, utilising zero-dimensional approaches, such as ordinary differential equations. With increases in computational power, models which incorporate space are increasingly used (e.g. partial differential equations, cellular automata). This inclusion of space permits dynamics not present in non-spatial frameworks, and illuminates processes that lead to pattern formation in ecological systems.

### Examples

One of the earliest, and most well-known, ecological models is the predator-prey model of Alfred J. Lotka (1925) and Vito Volterra (1926). This model takes the form of a pair of ordinary differential equations, one representing a prey species, the other its predator.

$$\frac{dX}{dt} = \alpha.X - \beta.X.Y$$

$$\frac{dY}{dt} = \gamma.\beta.X.Y - \delta.Y$$

where,

- $X$ is the number/concentration of the prey species;
- $\beta$ is the predation rate of $Y$ upon $X$;
- $Y$ is the number/concentration of the predator species;
- $\gamma$ is the assimilation efficiency of $Y$;
- $\alpha$ is the prey species' growth rate;
- $\delta$ is the mortality rate of the predator species

Volterra originally devised the model to explain fluctuations in fish and shark populations observed in the Adriatic Sea after the First World War (when fishing was curtailed). However, the equations have subsequently been applied more generally. Although simple, they illustrate some of the salient features of ecological models: modelled biological populations experience growth, interact with other populations (as either predators, prey or competitors) and suffer mortality.

Ecosystem models are mathematical representations of ecosystems. Typically they simplify complex foodwebs down to their major components or trophic levels, and quantify these as either numbers of organisms, biomass or the inventory/concentration of some pertinent chemical element (for instance, carbon or a nutrient species such as nitrogen or phosphorus).

### *Modelling of Infectious Disease*

It is possible to model the progress of most infectious diseases mathematically to discover the likely outcome of an epidemic or to help manage them by vaccination. This field tries to find parameters for various infectious diseases and to use those parameters to make useful calculations about the effects of a mass vaccination programme.

## Mathematical Modelling of Infectious Disease

It is possible to mathematically model the progress of most infectious diseases to discover the likely outcome of an epidemic or to help manage them by vaccination. This article uses some basic assumptions and some simple mathematics to find parameters for various infectious diseases and to use those parameters to make useful calculations about the effects of a mass vaccination programme.

### History

Early pioneers in infectious disease modelling were William Hamer and Ronald Ross, who in the early twentieth century applied the law of mass action to explain epidemic behaviour. Lowell Reed and Wade Hampton Frost developed the Reed-Frost epidemic model which described the relationship between susceptible, infected and immune individuals in a population.

### Concepts

$R_0$, the basic reproduction number :

The average number of other individuals each infected individual will infect in a population that has no immunity to the disease.

***S*** **:** The proportion of the population who are susceptible to the disease (neither immune nor infected).

***A:*** The average age at which the disease is contracted in a given population.

***L:*** The average life expectancy in a given population.

### Assumptions

- We assume a rectangular age distribution, such as that which is typically found in developed countries where there is a low infant mortality and much of the population lives to the life expectancy. In developed countries this assumption is often well justified.
- We also assume homogeneous mixing of the population, i.e., individuals of the population under scrutiny assort and make contact at random and do not mix mostly in a smaller subgroup. This assumption is rarely justified, when dealing with a country such as the UK, because most people in London, say, only make contact with other Londoners. If we deal only with London, then there will be smaller subgroups such as the Turkish community or teenagers (just to give two examples) who will mix with each other more than people outside their group. However, homogeneous mixing is a standard assumption to make the mathematics tractable.

Whenever we are modelling anything mathematically, whether in epidemiology or otherwise, we would be wise to remember that a mathematical model is only as good as the assumptions on which it is based.

If a model makes predictions which are out of line with observed results and the mathematics is correct, we must go back and change our initial assumptions in order to make the model useful.

### The Endemic Steady State

An infectious disease is said to be endemic when it can be sustained in a population without the need for external inputs. This means that, on average, each infected person is infecting *exactly* one other person (any more and the number of people infected will grow exponentially and there will be an epidemic, any less and the disease will die out). In mathematical terms, that is:

$$R_0 \times S = 1.$$

The basic reproduction number ($R_0$) of the disease, assuming everyone is susceptible, multiplied by the proportion of the population that is actually susceptible ($S$) must be one (since those who are not susceptible do not feature in our calculations as they cannot contract the disease). Notice that this relation means that for a disease to be in the endemic steady state, the higher the basic reproduction number, the lower the proportion of the population susceptible must be, and vice versa; a mathematical basis for a result that might have been intuitively obvious. The first assumption lets us say that everyone in the population lives to age $L$ and then dies. If the average age of infection is $A$, then on average individuals younger than $A$ are susceptible and those older than $A$ are immune (or infectious). Thus the proportion of the population that is susceptible is given by:

$$S = \frac{A}{L}.$$

But the mathematical definition of the endemic steady state can be rearranged to give:

$$S = \frac{1}{R_0}.$$

And therefore, since things equal to the same thing are equal to each other:

$$\frac{1}{R_0} = \frac{A}{L}.$$

$$R_0 = \frac{L}{A}.$$

This provides us with a simple way to estimate the parameter $R_0$ using easily available data.

### In a Population with an Exponential Age Distribution

For a population with an exponential age distribution, it turns out that;

$$R_0 = 1 + \frac{L}{A}.$$

The mathematics required to calculate this is a little more complicated than that above, and thus beyond the scope of this article. However, this does allow you to work out the basic reproduction number of a disease given $A$ and $L$ in either type of population distribution.

### Infectious Disease Dynamics

In recent years it has become obvious that there is a need to accommodate the growing integration of quantitative methods with the increasing volume of data being generated on host-pathogen interactions. This has resulted in a growing body of research covering quantitative or theoretical studies of the population dynamics, structure and evolution of infectious diseases of plants and animals, including humans.

Specific topics in this area include:

- transmission, spread and control of infection
- epidemiological networks
- spatial epidemiology
- persistence of pathogens within hosts
- intra-host dynamics
- immuno-epidemiology
- virulence
- Strain (biology) structure and interactions
- antigenic shift
- phylodynamics
- pathogen population genetics
- evolution and spread of resistance
- role of host genetic factors
- statistical and mathematical tools and innovations
- role and identification of infection reservoirs.

### The Mathematics of Mass Vaccination

If the proportion of the population that is immune exceeds the herd immunity level for the disease, then the disease can no longer persist in the population. Thus, if this level can be exceeded by vaccination, the disease can be eliminated. An example of this being successfully achieved worldwide is the global eradication of smallpox, with the last wild case in 1977. Currently, the WHO is carrying out a similar campaign of vaccination in an attempt to eradicate polio.

The herd immunity level will be denoted $q$. Recall that, for a stable state:

$$R_0.S = 1.$$

$S$ will be $(1 - q)$, since $q$ is the proportion of the population that is immune and $q + S$ must equal one (since in this simplified model, everyone is either susceptible or immune). Then:

$$R_0.(1 - q) = 1,$$

$$1 - q = \frac{1}{R_0},$$

$$q = 1 - \frac{1}{R_0}.$$

Remember that this is the threshold level. If the proportion of immune individuals *exceeds* this level due to a mass vaccination programme, the disease will die out. We have just calculated the critical immunisation threshold (denoted $q_c$). It is the minimum proportion of the population that must be immunised at birth (or close to birth) in order for the infection to die out in the population.

$$q_c = 1 - \frac{1}{R_0}$$

### When a Mass Vaccination Programme Cannot Exceed the Herd Immunity

If the vaccine used is insufficiently effective or the required coverage cannot be reached (for example due to popular resistance) the programme may not be able to exceed $q_c$. Such a programme can, however, disturb the balance of the infection without eliminating it, often causing unforeseen problems. Suppose that a proportion of the population $q$ (where $q < q_c$) is immunised at birth against an infection with $R_0 > 1$. The vaccination programme changes $R_0$ to $R_q$ where

$$Rq = R_0(1 - q)$$

This change occurs simply because there are now fewer susceptibles in the population who can be infected. $R_q$ is simply $R_0$ minus those that would normally be infected but that cannot be now since they are immune.

As a consequence of this lower basic reproduction number, the average age of infection $A$ will also change to some new value $A_q$ in those who have been left unvaccinated.

Recall the relation that linked $R_0$, $A$ and $L$. Assuming that life expectancy has not changed, now:

$$Rq = \frac{L}{A_q},$$

$$Aq = \frac{L}{R_q},$$

$$Aq = \frac{L}{R_0(1-q)}.$$

But $R_0$ = L/A so:

$$Aq = \frac{L}{(L/A)(1-q)},$$

$$Aq = \frac{AL}{L(1-q)},$$

$$Aq = \frac{A}{(1-q)}.$$

Thus the vaccination programme will produce an increase in the average age of infection, another mathematical justification for a result that might have been intuitively obvious. Unvaccinated individuals now experience a reduced force of infection due to the presence of the vaccinated group.

However, it is important to consider this effect when vaccinating against diseases which increase in severity with age. A vaccination programme against such a disease that does not exceed $q_c$ may cause more deaths and complications than there were before thc programme was brought into force as individuals will be catching the disease later in life. These unforeseen outcomes of a vaccination programme are called perverse effects.

## Epidemic Model

An Epidemic model is a simplified means of describing the transmission of communicable disease through individuals. The outbreak and spread of disease has been questioned and studied for many years. The ability to make predictions about diseases could enable scientists to evaluate inoculation or isolation plans and may have a significant effect on the mortality rate of a particular epidemic. The modelling of infectious diseases is a tool which has been used to study the mechanisms by which diseases spread, to predict the future course of an outbreak and to evaluate strategies to control an epidemic. The first scientist who systematically tried to quantify causes of death was John Graunt in his book *Natural and Political Observations made upon the Bills of Mortality*, in 1662. The bills he studied were listings of numbers and causes of deaths published weekly. Graunt's analysis of causes of death is considered the beginning of the "theory of competing risks" which according to Daley and Gani is "a theory that is now well established among modern epidemiologists".

The earliest account of mathematical modelling of spread of disease was carried out in 1766 by Daniel Bernoulli. Trained as a physician, Bernoulli created a mathematical model to defend the practice of inoculating against smallpox (Hethcote, 2000). The calculations from this model showed that universal inoculation against smallpox would increase the life expectancy from 26 years 7 months to 29 years 9 months. Following Bernoulli, other physicians contributed to modern mathematical epidemiology. Among the most acclaimed of these were A. G. McKendrick and W. O. Kermack, whose paper *A Contribution to the Mathematical Theory of Epidemics* was published in 1927. A simple deterministic (compartmental) model was formulated in this paper. The model was successful in predicting the Behaviour of outbreaks very similar to that observed in many recorded epidemics.

### Types of Epidemic Models

***Stochastic:*** "Stochastic" means being or having a random variable. A stochastic model is a tool for estimating probability distributions of potential outcomes by allowing for random variation in one or more inputs over time. Stochastic models depend on the chance variations in risk of exposure, disease and other illness dynamics. They are used when these fluctuations are important, as in small populations.

### Deterministic

When dealing with large populations, as in the case of tuberculosis, deterministic or compartmental mathematical models are used. In the deterministic model, individuals in the population are assigned to different subgroups or compartments, each representing a specific stage of the epidemic. Letters such as M, S, E, I, and R are often used to represent different stage. The transition rates from one class to another are mathematically expressed as derivatives, hence the model is formulated using differential equations. While building such models, it must be assumed that the population size in a compartment is differentiable with respect to time and that the epidemic process is deterministic. In other words, the changes in population of a compartment can be calculated using only the history used to develop the model.

Another approach is through discrete analysis on a lattice (such as a two-dimensional square grid), where the updating is done through asynchronous single-site updates (Kinetic Monte Carlo) or synchronous updating (Cellular Automata). The lattice approach enables inhomogeneities and clustering to be taken into account. Lattice systems are usually studied through computer simulation, and are discussed in the Wikipedia page Epidemic models on lattices.

### Terminology

The following is a summary of the notation used in this and the next sections.

- M — Passively Immune Infants
- S — Susceptibles
- E — Exposed Individuals in the Latent Period
- I — Infectives
- R — Removed with Immunity
- $\beta$ — Contact Rate
- $\mu$ — Average Death Rate
- B — Averate Birth Rate
- $1/\varepsilon$ — Average Latent Period
- $1/\gamma$ — Average Infectious Period
- $R_0$ — Basic Reproduction Number
- N — Total Population
- $f$ — Average Loss of Immunity Rate of Recovered Individuals
- $\delta$ — Average Temporary Immunity Period.

### Reproduction Number

There is a threshold quantity which determines whether an epidemic occurs or the disease simply dies out. This quantity is called the basic reproduction number, denoted by $R_0$, which can be defined as the number of secondary infections caused by a single infective introduced into a population made up entirely of susceptible individuals (S(0) H• N) over the course of the infection of this single infective. This infective individual makes $\beta$N contacts per unit time producing new infections with a mean infectious period of $1/\gamma$. Therefore, the basic reproduction number is

$$R_0 = (\beta N)/\gamma$$

This value quantifies the transmission potential of a disease. If the basic reproduction number falls below one ($R_0 < 1$), i.e. the infective may not pass the infection on during the infectious period, the infection dies out. If $R_0 > 1$ there is an epidemic in the population. In cases where $R_0 = 1$, the disease becomes endemic, meaning the disease remains in the population at a consistent rate, as one infected individual transmits the disease to one susceptible.

In cases of diseases with varying latent periods, the basic reproduction number can be calculated as the sum of the reproduction number for each transition time into the disease. An example of this is tuberculosis. Blower et al. (1995) calculated from a simple model of TB the following reproduction number:

$$R_0 = R_0^{\text{Fast}} + R_0^{\text{Slow}}$$

In their model, it is assumed that the infected individuals can develop active TB by either direct progression (the disease develops immediately after infection) considered above as FAST tuberculosis or endogenous reactivation (the disease develops years after the infection) considered above as SLOW tuberculosis.

## Other Considerations within Compartmental Epidemic Models

### *Vertical Transmission*

In the case of some diseases such as AIDS and Hepatitis B, it is possible for the offspring of infected parents to be born infected. This transmission of the disease down from the mother is called Vertical Transmission. The influx of additional members into the infected category can be considered within the model by including a fraction of the newborn members in the infected compartment.

### *Vector Transmission*

Diseases transmitted from human to human indirectly, i.e. malaria spread by way of mosquitoes, are transmitted through a vector. In these cases, the infection transfers from human to insect and an epidemic model must include both species, generally requiring many more compartments than a model for direct transmission. For more information on this type of model see the reference *Population Dynamics of Infectious Diseases: Theory and Applications*, by R. M. Anderson.

### *Others*

Other occurrences (taken from *Mathematical Models in Population Biology and Epidemiology* by Fred Brauer and Carlos Castillo-Chavez) which may need to be considered when modelling an epidemic include things such as the following:

- Nonhomogeneous mixing
- Age-Structured populations

- Variable infectivity
- Distributions that are spatially non-uniform
- Diseases caused by macroparasites
- Acquired immunity through vaccinations.

## Stochastic Simulation

Stochastic simulation algorithms and methods were initially developed to analyse chemical reactions involving large numbers of species with complex reaction kinetics. The first algorithm, the Gillespie algorithm was proposed by Dan Gillespie in 1977. It is an exact procedure for numerically simulating the time evolution of a well-stirred chemically reacting system.

## Gillespie Algorithm

In probability theory, the Gillespie algorithm (or occasionally the Doob-Gillespie algorithm) generates a statistically correct trajectory (possible solution) of a stochastic equation. It was created by Joseph L. Doob and others (circa 1945) and popularized by Dan Gillespie in 1977 in a paper where he uses it to simulate chemical or biochemical systems of reactions efficiently and accurately using limited computational power. As computers have become faster, the algorithm has been used to simulate increasingly complex systems. The algorithm is particularly useful for simulating reactions within cells where the number of reagents typically number in the tens of molecules (or less). Mathematically, it is a variety of a dynamic Monte Carlo method and similar to the kinetic Monte Carlo methods. It is used heavily in computational systems biology.

### *History*

The process that lead to the algorithm recognizes several important steps. In 1931, Andrei Kolmogorov introduced the differential equations corresponding to the time-evolution of stochastic processes that proceed by jumps, today known as Kolmogorov equations (Markov jump process) (a simplified version is known as master equation in the natural sciences). It was William Feller, in 1940, who found the conditions under which the Kolmogorov equations admitted (proper) probabilities as solutions. In his Theorem I (1940 work) he establishes that the time-to-the-next-jump was exponentially distributed and the probability of the next event is proportional to the rate. As such, he established the relation of Kolmogorov's equations with stochastic processes. Later, Doob (1942, 1945) extended Feller's solutions beyond the case of pure-jump processes. The method was implemented in computers by David

George Kendall (1950) using the Manchester Mark 1 computer and later used by M. S. Bartlett (1953) in his studies of epidemics outbreaks. Gillespie (1977) worked ignoring this history as he writes "It should be emphasized, though, that the master equation itself plays no role whatsoever in either the derivation or the implementation of the stochastic simulation algorithm". Gillespie then proceeds through an heuristic argument to introduce the algorithm.

### *Algorithm*

Gillespie developed two different, but equivalent formulations; the direct method and the first reaction method. Below is a summary of the steps to run the algorithm (math omitted):

1. Initialization: Initialize the number of molecules in the system, reactions constants, and random number generators.
2. Monte Carlo step: Generate random numbers to determine the next reaction to occur as well as the time interval. The probability of a given reaction to be chosen is proportional to the number of substrate molecules.
3. Update: Increase the time step by the randomly generated time in Step 2. Update the molecule count based on the reaction that occurred.
4. Iterate: Go back to Step 2 unless the number of reactants is zero or the simulation time has been exceeded.

The algorithm is computationally expensive and thus many modifications and adaptations exist, including the next reaction method (Gibson & Bruck), tau-leaping, as well as hybrid techniques where abundant reactants are modelled with deterministic Behaviour. Adapted techniques generally compromise the exactitude of the theory behind the algorithm as it connects to the Master equation, but offer reasonable realizations for greatly improved timescales. Recently, an exact version of the algorithm with constant-time scaling has been developed enabling efficient simulation of systems with very large numbers of reaction channels (Slepoy 2008).

# Bibliography


Abhilash M.: *Introduction to Bioinformatics and Microarray Technology,* CBS Pub, Delhi, 2010.

Agarwal, J.P.: *Introduction to Mathematical Methods in Bioinformatics,* Swastik Pub, Delhi, 2011.

Alam Imtiyaz: *Elementary Bioinformatics,* Pharma Book Syndicate, Delhi, 2005.

Bansal, Manju: *Basic Bioinformatics,* Atlantic Pub, Delhi, 2009.

Bosu, Orpita and Simminder Kaur Thukral: *Bioinformatics: Databases, Tools, and Algorithms* , Oxford University Press, Delhi, 2007.

Chavali, L.N.: *Bioinformatics and Bioprogramming in C,* Universities Press, Delhi, 2009.

Chowdhary, K.R. and V.S. Bansal: *Bioinformatics and Computational Technologies,* Scientific, Delhi, 2010.

Durbin, R.: *Biological Sequence Analysis: Probabilistic Models of Proteins and Nucleic Acids,* Cambridge University Press, Cambridge, 2004.

Fumento, Michael : *Bioevolution: How Biotechnology is Changing Our World,* San Francisco: Encounter, 2003.

Gaskell, G.: *Biotechnology: The Making of a Global Controversy,* Cambridge: Cambridge University Press, 2002.

Gray, I.C.: *Bioinformatics for Geneticists,* Wiley, 2003.

Harisha, S.: *Fundamentals of Bioinformatics,* I K International, Delhi, 2007.

Jean L.: *A Revolution in Biotechnology,* Cambridge, MA: Cambridge University Press, 1989.

Khan, Irfan Ali: *Emerging Trends in Bioinformatics,* Ukaaz, Delhi, 2002.

Khanum, Atiya: *Recent Advances in Bioinformatics,* Ukaaz Pub, 2003.

Lodish, Harvey: *Molecular Cell Biology,* New York: W. H. Freeman, 2000.

Lukas K.: *Bioinformatics Basics: Applications in Biological Science and Medicine,* Routledge, Delhi, 2010.

Manikandakumar, K.: *Dictionary of Bioinformatics,* MJP Pub, Delhi, 2009.

Niranjan Sarangi: *Applied Bioinformatics Statistics and Economics in Fisheries Research*, New India Pub Agency, Delhi, 2008.

Ouellette, B.F.F.: *Bioinformatics: A Practical Guide to the Analysis of Genes and Proteins*, Wiley, 2005.

Parens, Erik: *Enhancing Human Traits: Ethical and Social Implications*, Washington, DC: Georgetown University Press, 1998.

Pavel A Pevzner: *An Introduction to Bioinformatics Algorithms*, MIT Press, Delhi, 2004.

Prasad, S.K. : *Biochemistry of Proteins*, Discovery, Delhi, 2010.

Rastogi, S.C. : *Bioinformatics : Concepts, Skills & Applications*, CBS Pub, Delhi, 2003.

Rathi, K.P.: *Bioinformatics : Concepts, Elements and Dimensions*, Aadi, Delhi, 2010.

Roy, Ajit Kumar and Niranjan Sarangi: *Applied Bioinformatics Statistics and Economics in Fisheries Research,* New India Pub Agency, Delhi, 2008.

Shanmughavel, P.: *Principles of Bioinformatics,* Pointer, 2005.

Sharma, T.R.: *Genome Analysis and Bioinformatics : A Practical Approach*, I.K. International Publishing House, Delhi, 2009.

Subramanian, C.: *A Textbook of Bioinformatics*, Dominant, Delhi, 2004.

Suri, Nitin: *Molecular Biology and Biochemistry*, Oxford Book Company, Delhi, 2010.

Swarnim, K.: *A Textbook of Biochemistry and Microbiology*, Surendra Pub, Delhi, 2010.

Tannistha Nandi: *An Introduction to Bioinformatics*, Duckworth Press, Delhi, 2007.

Tripathi, Nirmal: *Molecular Biotechnology,* Crescent Publishing Corporation, Delhi, 2008.

Trivedi, Priya Ranjan ; Ravinder Singh Sohal and Haojam Rocky Singh : *Encyclopaedia of Bioinformatics* , Jnanada Prakashan, Delhi, 2007.

Vaishy, Ajay Kumar: *Modern Dictionary of Bioinformatics and Biotechnology,* Manglam, Delhi, 2008.

Wadhwa, Gulshan: *Practicals in Bioinformatics,* Pointer Pub, 2009.

Westhead, David R. ; J. Howard Parish and Richard M. Twyman: *Instant Notes : Bioinformatics,* Viva, London, 2010.